CW00553795

From
PAIN
to
POWER

"It offers an inspirational model for healing."
– Howard Schubiner, MD

From
PAIN
to
POWER

Break Free from Chronic Pain:
Your Roadmap to Healing

Narinder Sheena

FUZZY
FLAMINGO

First published in 2024 by Fuzzy Flamingo
Copyright © Narinder Sheena 2024

Narinder Sheena has asserted her right to be identified as the author of this
Work in accordance with the Copyright, Designs and Patents Act 1988.

ISBN: 978-1-7390943-5-5

All rights reserved.
No part of this publication may be reproduced, stored in a retrieval system,
or transmitted in any form or by any means, electronic, mechanical,
photocopying, recording or otherwise, without the prior permission of the
copyright owner.

Typeset by Fuzzy Flamingo
www.fuzzyflamingo.co.uk

A catalogue for this book is available from the British Library.

This book is dedicated to anyone suffering from a chronic condition right now.

Because of the dynamic nature of the Internet, any web addresses or links contained in this book may have changed since publication and may no longer be valid. The views expressed in this work are solely those of the author.

The author of this book does not dispense medical advice or prescribe the use of any technique as a form of treatment for physical, emotional, or medical problems without the advice of a physician, either directly or indirectly. The intent of the author is only to offer information of a general nature to help you in your quest for emotional and spiritual well-being. In the event you use any of the information in this book for yourself, which is your constitutional right, the author and the publisher assume no responsibility for your actions.

Contents

Foreword

Coming across the work of Dr John E. Sarno in 2007 was life-changing for me. Not only did it help me improve my health and wellbeing, but it completely changed the way I work and ultimately, my whole career.

As a physiotherapist, like many others trying to help people with chronic pain, I found it pretty frustrating because of the widely accepted understanding of the biomedical model. This was rooted in the belief that the cause of chronic pain was physical/structural, and therefore, all we could do was help them manage their pain/symptom. Even as understanding evolved to the awareness that chronic pain needed a biopsychosocial approach, this was still based on the belief that there was a physical cause for the pain. Beliefs are powerful, and this, reinforced by well-meaning health professionals, means an individual's progress will be limited to just learning to live with the pain.

As I became more and more interested in helping

people with pain, I decided to leave the National Health Service (NHS) and set up a private practice. This allowed me to provide more time for assessments and follow-up appointments with my patients. Sadly, many of them would say that this was the first time anyone had actually listened to them.

I found that my results improved with more time for communication as I also treated them with the gentle therapies I was using. I also noticed several anomalies that just didn't make sense if their pain was due to a physical cause, such as structural changes. This included:

- Pain being triggered and persisting despite the lack of any actual physical injury. Pain onset was often while doing something innocuous, like picking up a pen or getting out of bed. Frequently, it was while doing something they usually did without any problem.
- Many of my patients presented with pain that initially started on waking up one morning. They blamed it on their mattress or pillows, despite the fact they had slept on them for months or even years without a problem. Could these cause tissue damage?
- Pain resolving completely, even though surgery had been recommended for structural/ degenerative changes found on MRI scans – for example, prolapsed discs, facet joint disease, and

stenosis. There was no way my gentle treatments could change the structures of their spines.

After reading widely and speaking to various colleagues, I began to realise that pain was often triggered during or soon after a stressful time. I experienced this personally when I developed horrendous sciatica as I was getting out of bed one morning, soon after leaving the NHS and setting up my practice. This persisted until I saw a therapist who, for the first time in my experience, asked me what was going on in my life. I then became aware of how anxious I was about making sure I made a success of my private work, especially as my husband is self-employed and we had two teenage children. Outwardly, there was little indication of this because I was ignoring how I felt and trying to focus on building my business. It seems that this was not the best approach. When I reached a point of overwhelm, it manifested in pain, forcing me to step back a bit.

This understanding, and the fact that the pain resolved pretty quickly once the underlying cause was brought to my conscious awareness and addressed, started me on the track of learning more about how stress affects our health. Ultimately, this resulted in my coming across the work of Dr Sarno in 2007. Narinder will explain more about Dr Sarno later, but at this point, it was like an epiphany for me. I read all his books and began to buy one of his books in bulk in order to pass them on to any patients who were keen to learn about this concept and approach.

The results I saw personally and with my patients made it very clear that this was not something I could ignore. I was absolutely certain that Dr Sarno's findings since the 1970s were the missing links which could help many of my patients resolve their persistent pain – rather than just manage it – as well as prevent pain from recurring during times of stress.

Having found no one who knew about Dr Sarno in the United Kingdom or Europe, I contacted Dr Sarno, who invited me to New York to shadow him in his work, which I did in November 2007. With his support, on my return, I set up my patient programme based on his approach. At this time, some doctors in the United States had become interested in getting together to discuss how, as a group, we could raise awareness of Dr Sarno's work. Thankfully, I was able to return to the United States to attend this first meeting of health professionals in this field in 2009, followed by conferences there in 2010 and 2012. Since then, we have been running conferences in the United Kingdom.

During this time, I was feeling very isolated because all my peer support was in the United States. I realised that working on my own, it would take a long time to spread the word and have it accepted in the United

Kingdom. However, if I started to train other health professionals, this would raise awareness more quickly. Not only that, but I would gain peer support for myself in the United Kingdom and Europe. Plus, there would be trained practitioners to refer people to as required.

I therefore set up SIRPA, our training organisation, in 2010. SIRPA's was the first training developed for health professionals worldwide. We moved the training online in 2018 to make it more widely accessible. In 2012, I also developed an online recovery programme the public could access and work through themselves, as Narinder did.

Although Dr Sarno died in 2017, at nearly ninety-four years old, the work he pioneered for decades with little support continues to make a massive difference in people's lives. This is through his books, as well as the numerous health professionals and coaches who are now passionate about making a difference, as he did. There is a rapidly growing global community who are working hard to keep Dr Sarno's memory alive, as well as ensuring that the work continues to evolve and be evidence-based through ongoing studies.

Thankfully, pain science, as well as trauma science, is catching up. There are now hundreds of studies demonstrating the need for every health professional to be stress-aware and trauma-informed when seeing clients with chronic pain and other persistent conditions. Slowly but surely, we continue to raise awareness of the impact repressed emotions from the past, as well as unresolved emotions day to day, have on our health and wellbeing and how to address this. Following a medical assessment to rule out any possible tissue-damaging condition, this can result in a powerful and transformative approach for many people who have often struggled for years because the focus has been on the symptoms rather than the underlying causes.

Narinder is just one of tens of thousands of people worldwide who have benefited from the legacy Dr Sarno has left. As a result, Narinder is passionate about using her personal experiences and the understanding she has gained to help others through this book.

She provides hope through her own recovery story as well as numerous suggestions, strategies, and resources to provide readers with a powerful jump-start to their recoveries. For some, reading her book might be all you need to resolve your pain and regain your life. Others might benefit from some support and guidance, and with Narinder's background in coaching, I have no doubt she will continue to help more and more people.

Georgie Oldfield MCSP
Physiotherapist and Founder of SIRPA

Introduction

*"Don't get lost in your pain, know that one day
your pain will become your cure"*
– Rumi

If you suffer from chronic pain and you've found your way here, you probably have exhausted the medical system in your search to recover your healthy self, the person you remember before the pain. It's not only a physical issue, it's also about your mental well-being. The power and control you once had has slipped away and you want it back.

We often mistakenly believe chronic pain to be permanent and dangerous, leading us to think that we have no choice but to manage or live with it. We have no awareness or knowledge of it being anything different. However, it is important not to fear it. The fear of your body being broken through medical imaging, or the fear of words used to diagnose your pain, unfortunately causes our anxiety and nervous system to go on high alert, leaving us in a state of panic.

Imagine that you were waiting for surgery for your hip pain, awaiting a knee replacement, or a steroid injection for your shoulder. This means months of uncertainty enduring fear and anxiety, and then a compassionate healthcare professional steps in, offering another way out – one that sidesteps any invasive procedure. Would you take it?

Not all medical professionals have caught up with the holistic approach of dealing with individuals as a *whole person*, psychologically as well as physically. They are unaware that each individual is unique and has different coping strategies for such levels of pain. Fortunately, there is hope for change as we have access to extensive literature and evidence in the field of neuroscience. While it's natural for us to trust and rely on medical professionals, it is crucial to first understand the root cause of our ongoing pain ourselves. I am guessing you picked up this book to gain a deeper understanding and awareness of your pain's causes.

Your Choice

You probably didn't even know that there was a holistic approach to healing from chronic pain. You're taking an important step towards your healing and wanting to become pain-free. It's understandable that you've reached a point of exhaustion with the medical system. You may have spent a significant amount of money in search of long-term pain relief that has yet to materialise.

If you truly want to find a way to become pain-free, *understanding* how you have gotten into a chronic pain state in the first place will help you discover the root cause. Flicking through the chapters of this book, you will realise that it could easily have started in your childhood, your environment, or even your lived experiences. This could be through what we have witnessed or heard right up until the present moment. Our mindset, which takes into account the thoughts, feelings, and beliefs we absorb into our conscious and unconscious brains, plays a huge role in chronic pain.

This book is your personalised toolkit. It can be a constant source of inspiration and hope when you are struggling with your pain or the lack of self-belief that you can heal yourself through consistency and resilience. There are valuable insights included, even from a chronic pain sufferer who didn't think it was possible to become pain-free after thirty years by applying the knowledge they gained.

All the resources, useful tips, and exercises have been collected from my experience with chronic pain and my extensive knowledge gained through coaching, which helped me to heal. I had to find the courage to search through the available resources. These were books, reaching out to medical experts to support the evidence I was reading, extensive literature from articles based on neuroscience, and watching videos of real-life people healing. At the time, I could find no signposting or awareness of how I could heal in a holistic way. I hope this book can give you the signposting, information, and motivation to take that first step on your healing journey to becoming pain-free.

My Pain Journey

I vividly remember the excruciating pain that engulfed me, sweat cascading down my skin as I mustered all my strength to transition from a seated position to standing upright, albeit with a stooped posture. With desperate hands clutching onto the bathroom sink, I fought for stability as my body endured an onslaught of unbearable sensations, each one akin to a relentless knife piercing through me. The assault of pain and discomfort reverberated through every fibre of my being, stealing away any sense of respite or relief. As tears streamed down my face, a profound sense of powerlessness washed over me. Whilst the children were out of view, playing away and oblivious to my struggles, all I yearned for in that moment was to complete the simplest of tasks – brushing my teeth. It seemed inconceivable to me that such a mundane activity, once effortless, now loomed before me as an insurmountable challenge. The weight of frustration and helplessness bore down on my shoulders, amplifying the significance of this seemingly simple task.

The Pandemic

Several months had flown by, and I unexpectedly found myself transfixed by a headline about the relentless grip of coronavirus-19. This invisible enemy was claiming lives without rhyme or reason, leaving us all fearful. As a natural empath, I normally steered clear of the news, aware of its negative impact and the helplessness it stirred within me. The injustice of it all seared through my thoughts. *How bloody unfair*, I thought, and I could sense it was heading closer and closer to home. I was picturing a slow tsunami that would take us all by surprise.

Subconsciously, anxiety and fear took over, which added to the grief and suppressed emotions already residing within me. The pandemic acted as a potential catalyst in keeping me in a state of unease and apprehension. We were left wondering, *Will we be next?*

Flashback

It was a wintry February morning in 2020, just before

the world fell into the grip of lockdown. I rolled out of bed, exhausted from yet another restless night with my seven-month-old baby. It was then that I felt a peculiar twinge, a persistent ache in my right ankle. Days turned into weeks, and the pain lingered, gnawing at my thoughts like a persistent worry.

I had a flashback of a heart-stopping moment in which my worst fears of giving birth in my car became a stark reality. Yet, amidst the chaos and adrenaline, the midwives swooped in to catch my baby in time, ensuring that he arrived safely into this world. Relief washed over me as I held my baby in my arms.

I questioned what could have possibly caused this pain I was feeling in my body, unaware of any triggering event at that time. I longed for relief. I was *desperate* to find a solution.

Imagine physiotherapy over Zoom, where physical touch or that human connection is lost. That's how I felt as I was recommended an exercise to do that exacerbated the pain into the right lower area of my back to an excruciatingly painful level. This made childbirth feel like a walk in the park. If I could give you a visual of the pain, it would be somewhere between stubbing your little toe to hair being stripped off with hot wax for the first time. All I could feel was that burning, tingling, and stabbing sensation in my back, and I thought my pain threshold was high.

Fear

This was the moment I felt frightened, and it left me in a state of shock. I asked myself, what's happened to me? The lack of sleep and the constant nasty pain I kept feeling in different parts of my body left me wiped out. It was an alien kind of pain, not like the muscular pain you get when you have completed a half-marathon. I knew that type of pain. Having a high pain threshold, I thought I could cope, but I was wrong. I was on a merry-go-round full of fear and anxiety, going from one medical professional to the next, searching for hope. I remembered the moment when one of the top neurosurgeons in the country categorically advised me that I needed surgery. I was expecting hope, an alternative, something more. However, I was left feeling disappointed.

Acknowledging the Role of Accountability

When confronted with chronic pain and the seemingly irreversible limitations it imposes on us, we accept our roles in addressing our wellbeing. If we can acknowledge that we are both accountable and capable, this empowers us to take ownership of our situations. It is essential to let go of the idea that there is nothing more that can be done for us as this belief system can sabotage our progress. Instead, we allow a shift in our mindsets for all the possibilities we have yet to explore.

Challenging the Influences of Our Environment

The environment shapes our beliefs and sometimes limits our abilities to regain control of our lives. Society can condition us to believe that we need to passively accept our circumstances, leaving us feeling disempowered. However, it is important to understand that the environment neither defines nor dictates one's potential for growth. By questioning the narrative imposed upon us, we can free ourselves from these limitations and redirect our focus towards taking proactive steps to reclaim our *power* and focus on *healing*.

Embracing a Can-Do Attitude

Believing in our capabilities, strength, and resilience is crucial in the journey towards taking back control. We are not helpless; we can believe in ourselves that we can make a difference. It is through this mental shift that we can face the challenges ahead with unwavering determination. By embracing a can-do attitude, we unlock the potential within ourselves to explore new approaches to healing from chronic pain and ultimately become pain-free. Let us shatter the illusion that there is nothing more we can do for it is in these moments that we discover our true potential for growth and hope.

According to the World Health Organization in 2019, approximately 970 million individuals worldwide battled mental disorders, equating to 1 in 8 people across the globe. Anxiety experienced a staggering rise of over 25 per cent during that time.[1]

Additionally, in 2020, 619 million individuals worldwide grappled with the debilitating burden of low back pain, the leading cause of chronic pain. Unfortunately, these numbers are only expected to escalate in the future. But we can do something about it right now.[2]

---

1   See: https://www.who.int/news-room/fact-sheets/detail/mental-disorders.
2   See: https://www.who.int/news-room/fact-sheets/detail/low-back-pain.

CHAPTER 1

Regaining Personal Power: Awareness and Understanding

Pain is inevitable. Suffering is optional.
– Haruki Murakami

Trust me, I was you. And out of desperation, I reached out in so many directions my head hurt – the irony. I was scared and had that panicked feeling you get when you don't know how it would feel to sleep soundly on a comfortable bed, pick up your children, or even simply walk to the shops again. The sweat, tears, and pain are worth it. My struggle was my greatest strength, and it empowered me to want to help others like you to fight back with body blows, just like Rocky Balboa.

When was the last time we paused to reflect on ourselves or wholeheartedly directed our attention towards our families? In a world constantly fixated on *doing* and achieving, we often overlook the importance of simply *being* present. This begs the question: Why did it take a worldwide pandemic to prompt us to reconsider our mindsets and prioritise personal wellbeing more than ever before?

Dr Sarno

For me, it all started with gaining awareness and seeking knowledge. I came across Dr John Sarno's book, *Healing Back Pain,* and a light bulb went off in my mind. Dr Sarno enlightened me about the importance of treating the whole person, addressing not just the physical symptoms but also the psychological aspect. We long to be *heard,* to know that our doctors genuinely care when they see us. We want them to acknowledge the challenges we've faced throughout our lives. We refuse to be reduced to mere numbers or statistics as if our struggles are just another routine process in a busy medical practice. How can we bridge this gap between doctor and patient, ensuring that our voices are heard, our concerns acknowledged, and our unique stories understood?

When I spoke to Gemma McFall, a coach and pain specialist who overcame a decade of back pain, she shared her insights: "For me the psychoeducation was key. Without understanding why I was trapped in a pain-fear cycle I would never have been able to start on my journey to recovery. From this point forward it was just about finding ways to be in a state of flow doing things I love with the people I love."[1]

Acute vs Chronic Pain

Chronic pain often arises after three to six months, persisting even after the initial injury or illness has healed. The pain signals remain active in the nervous system for extended periods, resurfacing with intensified symptoms. Did an injury trigger your chronic pain? Interestingly, my own experience with chronic pain unfolded gradually, without a specific injury or illness. Looking back, I realised the immense toll my body had endured, from the strain of delivering my baby in the car to the months of isolation during lockdown. The absence of my loved ones and the absence of work led to the demise of my emotional well-being, leaving me with feelings of fear, anxiety, and sadness. It all began with a persistent ache that demanded my attention, which I ignored, but that was just the tip of the iceberg.

Surprisingly, chronic pain is more common than we realise, particularly among women, who tend to experience it more frequently than men, with issues like low back pain and migraines, two conditions I struggled with before finding a way out. When we focus on pain, we fixate on the physical sensations

---

1   Gemma McFall, Gallup strengths coach.

rather than considering the emotional aspects that are consuming us. Often, we fail to connect the dots that tie chronic pain to recurring childhood triggers that manifest in various forms throughout our lives.

According to an article published in June 2021 by the National Library of Medicine (NLM), women are reported to experience higher levels of chronic pain, especially when it comes to migraines, rheumatological issues, musculoskeletal pain, and fibromyalgia. Additionally, the menstrual and reproductive phases of a woman's life can have an impact on pain levels as well.[2]

The purpose of pain is to move us into action. It is not to make us suffer.
– Tony Robbins

Acknowledging Pain Sensations

The preceding quote from Tony Robbins served as a constant reminder of my struggles. It became clear to me that connecting the mind and body is crucial in understanding the triggers behind chronic pain and learning to acknowledge the sensations when they arise within us. We must ask ourselves, "What message is the pain trying to tell me?" The onset of pain in our bodies serves as a signal that something

---

2   See: https://www.ncbi.nlm.nih.gov/pmc/articles/PMC8119594/

is amiss. Dr Sarno, a renowned physician, achieved recognition through his groundbreaking work on the mind-body connection. He termed this phenomenon "tension myositis syndrome" (TMS). This became myoneural in his latter books which refers to the tension that manifests in our muscles, nerves, tendons, and ligaments often starting from childhood. It's an emotional journey intertwined with our physical wellbeing.

As I delved further into Dr Sarno's book, I found it to be a refreshing and effortlessly comprehensible guide. The daily affirmations, such as, "TMS is a harmless condition caused by my repressed emotions", and, "physical activity is not dangerous", reinforced the belief that the various sensations I experienced were safe and not threatening. Whenever I had doubts about the pain I was feeling, I would return to the reminders for reassurance.

It's important to recognise that acute injuries typically occur suddenly and can lead to fractures, bruises, or muscle tears. The recovery time for these injuries can vary from a few weeks to several months, depending on the severity. In some cases, complex surgeries and rehabilitation may extend the recovery period. The good news is that acute injuries are often temporary and can be healed with medical intervention, rest, and proper rehabilitation.

Mind-Body Connection

As I began to explore the mind-body connection, I discovered a significant link between anger, frustration, and muscle tension in the body. It surprised me to realise that anger was lurking in my subconscious, or unconscious brain, hidden away without my conscious awareness. The key lies in accepting and believing that the pain originates from a psychological place in our mind. This aligns with the concept of conversion disorder as initially described by Freud, the Austrian neurologist and founder of psychoanalysis, many years ago. Emotions find their way into physical manifestations, emphasising the profound influence of our psychological state on our physical wellbeing.[3]

We've all heard the dismissive phrase, "It's all in your mind." I recall a family member once saying this to me, and although I maintained a smile on the outside, I was boiling with frustration internally. It felt as though this person had no understanding of the genuine pain I was feeling and the struggles I was going through. We need to grasp how we perceive pain and what processes occur in the brain to generate it. This pain

---

3   See: https://www.ncbi.nlm.nih.gov/pmc/articles/
     PMC4479361/#:~:text=The%20term%20conversion%20
     disorder%20was,organic%20diseases%20reflect%20
     unconscious%20conflict.&text=The%20word%20conversion%-
     20refers%20to,symptom%20for%20a%20repressed%20idea.

is undeniably real, far from being exaggerated as some might suggest, so even after the initial injury has healed (acute), the pain persists, which is why it becomes chronic, and is generated by the brain.

Hope to Heal

After devouring Dr Sarno's book, I was captivated by the prospect of delving deeper into the realm of mind-body suffering. That's when I discovered Georgie Oldfield, a physiotherapist who wholeheartedly embraced Dr Sarno's principles. Even amidst my ongoing physiotherapy and chiropractic treatments, she instilled unwavering belief in my ability to recover, even when I lost belief in myself.

Those were the words I yearned to hear, the glimmer of hope my weary brain desperately needed. I told myself that I could conquer my chronic pain, one small step at a time, bolstered by newfound confidence. No doubt remained that my distressing symptoms were stress-induced, originating solely within my mind. It was a relief to have someone identify the true cause. Yet, a nagging doubt lingered in the depths of my consciousness, wondering if there might be a serious underlying issue. The intensity and sheer unpleasantness of the pain fed that doubt. Determined, I embarked on the online recovery programme Georgie developed for SIRPA at my own pace, wanting to uncover how I unconsciously created my pain. She became a beacon of strength amidst my incessant self-doubt, fear, and anxiety. Her supportive words and extensive knowledge provided immense comfort, reassuring me that she was right beside me every step of the way.

In order to empathize with someone's experience you must be willing to believe them as they see it and not how you imagine their experience to be.
– Brené Brown

Belief

I would always recommend that you seek medical clearance to rule out any potential cancerous growth, fracture, or infection when dealing with chronic pain. It's important to ensure that there are no underlying serious medical conditions causing your symptoms. Georgie emphasised the importance of working on my mindset rather than seeking all forms of medical treatment. She encouraged me to understand that relying on external interventions suggested a need for someone or something else to fix me, whereas true healing came from within by doing the inner work. Blending medical treatments with a mind-focused approach could muddy the waters and keep the pain symptoms lingering due to fear and confusion. Seeking medical advice had already left me in a state of confusion. It's natural for this kind of confusion to breed anxiety and fear of the unknown, which only solidifies the grip of pain.

The belief we must hold, without any shred of doubt, is that there exists a profound mind-body connection. However, I laugh because acknowledging and accepting this truth is never as simple as it sounds. Our ingrained beliefs, which can be stubbornly difficult to shift, paint a picture where medical professionals are the only trustworthy authority figures. Undoubtedly, they possess extensive training and perform commendable work.

Nonetheless, did they enquire about the circumstances of your life when the pain arose? Were they limited by time constraints, barely able to scratch the surface beyond listing your symptoms? Perhaps medications or further diagnostic tests like X-rays or scans were necessary. Yet throughout the endless waiting periods between appointments, did it truly relax your mind or cause a lingering sense of fear and constant pain?

Exercise

1. Notice if you are feeling increased pain on a particular day, and jot down what may have caused it. Ask yourself what may have influenced the increased pain.

2. If you are suffering from multiple issues in your body, can you focus on how you are feeling that day? What about just writing that feeling down?
3. As feelings of anger, frustration, and sadness will inevitably be with you, how can you be kind to yourself?
4. Check in with your feelings daily. When you wake up, how are you feeling? Notice throughout the day if a strong emotion comes over you, and track it to see if a pattern emerges.

I read an Instagram post by Dr Will Cole that read as follows:

Dear Body,

Thank you for looking after me even when I wasn't looking after you. I'm sorry for not listening when you whispered to me in subtle signs.

I'm sorry for resenting you when you screamed in symptoms.

I promise to listen to you, to nourish you, and to take care of you the way you have taken care of me.

Can you write something similar regarding your pain symptoms as a reminder?

(@drwillcole) Instagram

People-Pleasing

Dr Sarno, a groundbreaking pioneer ahead of his time, initially discovered something remarkable while working as a physician in the field of rehabilitation medicine. He observed that patients attending his clinic experienced significant improvements from simply engaging in conversations, without the need for any physical therapy. Intrigued by this phenomenon, Dr Sarno started to recognise that these individuals benefited from speaking about stressful life events, which made him curious.

Unfortunately, his articles, evidence-based work emphasising the importance of a holistic approach to health care, were largely overlooked by much of the medical industry at that time. As the years passed, medical experts in the mind-body field have substantiated his findings repeatedly, highlighting the role an individual's psychology – as well as the intricate connections between the brain, nervous system, and gut – plays in understanding chronic pain. By acknowledging both psychological and physiological factors, healthcare professionals can successfully facilitate substantial pain reduction for their patients.

In an article published in *Psychology Today* (April 2023), titled "Smiling to Death: The Hidden Dangers of Being Nice," Gabor Mate, a physician and author of *When the Body Says No*, observed a correlation between chronic illness and specific personality traits. These traits include prioritising the emotional needs of others while neglecting one's own, suppressing anger, and maintaining a constant desire to please others. These characteristics may resonate with you and me.[4]

Dr Sarno, drawing from his extensive experience working with patients, made a similar observation. He noticed that individuals who exhibited certain personality traits, such as people-pleasing, subconscious anger, and perfectionism, were more susceptible to chronic pain. Through education and increasing self-awareness, Dr Sarno aimed to help these individuals understand the root cause behind their chronic pain.

I am not what happened to me,
I am what I choose to become.
– Carl Jung

Adverse Childhood Experiences

Adverse childhood experiences (ACEs) can unlock a torrent of pain and torment upon innocent young

---

4   See: https://www.psychologytoday.com/gb/blog/its-not-you-
     its-the-world/202304/smiling-to-death-the-hidden-dangers-of-
     being-nice

souls, especially those between birth and eighteen years old. Within this harrowing realm, a multitude of heart-wrenching situations prevails:

- The cruel touch of abuse, striking with emotional, physical, and sexual scars.
- The chilling void of neglect leaves young hearts devoid of emotional nourishment and physical care.
- The chaos inside dysfunctional homes, reveals the pain of separation or divorce between parents to mental illness, substance abuse, violence, and even death.

I plunged deep into the abyss of my past and the memory of my mother's nervous breakdown unfolding in front of my small eyes. It led me to question whether witnessing such an event left me scarred with hidden trauma.

What if all those years ago I had mustered the courage to confide in someone, to release this burden from the depths of my mind that has remained untouched until now?

As a child growing up in the eighties, shame, embarrassment, and ridicule loomed large. The topic of mental illness, suffocated by silence, was not afforded the open discussion it so desperately deserved. We were left to navigate the treacherous terrain of mental illness, grappling with unspoken anguish in the years before mental wellbeing was advocated so passionately worldwide. Healing and understanding can thrive when there is open discussion.

Anxiety

As individuals with anxiety, we become adept at concealing this aspect of ourselves, skilfully presenting a mask when necessary, be it in the workplace or other demanding environments. People often remark on our seemingly unshakable composure, and outwardly, this holds true. From a young age, I discovered coping strategies to help overcome the internal anxiety bubbling away to the surface. I now question, as an empath, how does one rinse the sponge that has absorbed an overwhelming surge of emotions over the years when no one else is aware?

So many of us can relate to that longing, yearning for someone to truly *see* through the façade we put up and acknowledge the fears and struggles we grapple with, just like everyone else. It's a natural tendency to present ourselves as confident, organised, and always ready to lend a helping hand. Yet beneath the surface, there lies an emotional struggle, a constant battle against our feelings of inadequacy and an array of fears that threaten to overwhelm us.

> *Waiting is painful. Forgetting is painful. But not knowing which to do is the worse kind of suffering.*
> – Paolo Coelho

According to a compelling article published by Jim Folk on anxietycentre.com, chronic pain serves as a prevalent anxiety disorder, capable of exacerbating other pain-related conditions including arthritis, irritable bowel syndrome (IBS), and fibromyalgia. When anxiety takes hold, it triggers a cascade of events, activating the stress response and heightening sensitivity and reactivity to pain. This dysregulation within the nervous system can even lead to neuropathic and phantom pain, further intensifying the pain.[5]

---

5   See: https://www.anxietycentre.com/anxiety-disorders/symptoms/chronic-pain/

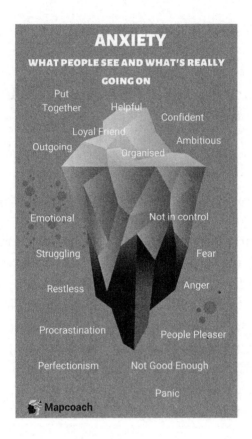

Desperately Seeking a Solution

When desperation grips us, we often turn to medications, hoping for respite from our suffering. Yet, these remedies come with their own set of potential side effects – weight gain, disrupted sleep patterns, and mood fluctuations. Even the most effective ibuprofen failed to alleviate the pain that consumed me. Desperation drove me to my local doctor, who prescribed painkillers awaiting collection at the pharmacy. But doubts kept niggling at me. Could I develop tolerance to these painkillers? I made a choice: I left them behind and turned inward, asking myself what the pain was trying to tell me.

Medication can offer short-term relief for chronic pain, but how long until we crave more of the same as the pain resurfaces? In the United States, an opioid crisis has been prevalent for a number of years. In the UK there has been a significant increase, despite guidelines from the National Institute for Health and Care Excellence (NICE) advising against opioid use for chronic primary pain since April 2021.

Our struggles remain hidden, unheard, and unseen. Isolated, we battle unimaginable pain, yearning to convey our truth to others. And in our vulnerability, we question whether they truly care.[6]

Imagine a blissful moment beneath the soothing stream of a shower, feeling the warmth all over your body while the weight of standing becomes a forgotten burden. Just one precious minute to allow your thoughts to drift towards your aspirations for the day, exploring new ideas rather than being consumed by pain.

Picture yourself strolling effortlessly to the nearby

---

6   https://www.ncbi.nlm.nih.gov/pmc/articles/PMC10278447/.

shops without the need to stop every few moments, gasping for breath as the pain drains your energy. It might seem as though you've endured a marathon when, in reality, you've only taken a few steps. But there's no panting or exhaustion in sight. How incredible would it be to continue walking further than ever before, even knowing that pain may strike again? With unwavering determination and boundless passion, you'd greet the challenge fearlessly, knowing that you don't give up easily.

If we can learn to not fear chronic pain, especially when it comes and goes, shifting its presence within our bodies, this will help reinforce the belief that we are experiencing a mind-body connection. It's only natural for us to desperately seek relief as we are consumed by the desire to escape from this excruciating agony. Quick fixes from medical professionals or medication become our sole focus, all in pursuit of temporary respite. However, what if we can *embrace this pain* as a signal of deeper psychological triggers and accept its presence? Treat it as a companion, one that holds a purpose for us, urging us to make necessary changes in our lives.

Dr Sarno recognised that the physical pain we so desperately try to avoid serves as a distraction from the emotional pain we are reluctant to confront. The path towards healing our emotional wounds, those hidden sources of unconscious chaos within our bodies, is undoubtedly uncomfortable. The thought of facing it may be daunting. But with a heightened awareness of the inner workings of our minds, a glimmer of hope illuminates our paths. It reminds us that we, too, can step into the person we aspire to be, especially when we consider the reasons driving our healing journey.

Exercise

Create a timeline on a piece of paper, mapping out your journey from childhood to the present moment. Recall any significant stressful triggers or traumatic events that remain etched in your memory which may have contributed to your experience of chronic pain. These could manifest as tension headaches, migraines, or back pain, just to name a few. Even seemingly innocent comments made by others, whether in childhood or adulthood, that consistently affect you deserve to be written on this timeline.

If you have endured adverse childhood experiences (ACEs), take a moment to reflect on whether you identify as a people-pleasing perfectionist. If the answer is yes, consider what changes you can implement to reclaim your power. Setting boundaries is one example.

Consider whether certain individuals, or situations no longer serve your wellbeing, and take proactive steps to distance yourself from them.

I understand that implementing boundaries can be particularly challenging within a family environment, where a sense of obligation may seem inescapable. However, it is important to recognise that your chronic pain is a result of the unconscious conflict you experience between your authentic desires and the expectations placed upon you. Prioritising your wellbeing is your ultimate goal.

CHAPTER 2

What Does Neuroscience Have to Do with Chronic Pain?

Knowledge is Power.
– Francis Bacon

As I continued searching for a solution to my chronic pain, I embarked on a journey of self-education and empowerment. It began with my discovery of Dr Sarno's book, which opened my eyes to new possibilities. This exploration led me down the fascinating path of neuroscience, human behaviour, and a deeper understanding of my own brain's functions. I became intrigued by the intricate interplay of electrical and chemical signals within our brains as they process information.

In May 2023, Ilona Garner, a researcher in neurological surgery, published an article detailing a small-scale study conducted in America. The study aimed to explore how different regions of the brain, in particular the anterior cingulate cortex (ACC) and orbitofrontal cortex (OFC), responded to heat and distinguished between acute and chronic pain. The findings revealed that the neural activity relating to pain came from signals generated in the ACC and didn't last as long as the sustained signals in the OFC. While this research represents just the tip of the iceberg, it highlights the pressing need for a personalised approach to caring for each individual struggling with chronic pain. Rewiring the neural circuits of our brains becomes imperative for finding sustainable relief.[7]

Awareness and Understanding

My journey towards understanding my experiences took a transformative turn when I reached out to Georgie Oldfield. Through her guidance, I began to unravel the true nature of my condition. I discovered that neuroscience holds the key to comprehending how the human brain functions and impacts our thoughts, emotions, behaviours, and physical sensations. Just like a meticulously programmed computer, our brains control the body's systems. However, as with any technology, there are bound to be glitches along the way. Our operating patterns are influenced by many factors – beliefs, environments, and the unique lens through which we perceive the

---

7   See: https://www.ucsf.edu/news/2023/05/425386/has-science-cracked-code-chronic-pain.

world. Once we start understanding this complex interplay of factors, we can start to navigate our journey towards healing.

Healing doesn't mean the damage never existed. It means the damage no longer controls our lives.
– Akshay Dubey

Amygdala

The amygdala is a small, almond-shaped structure in the brain and part of the limbic system. It's primarily associated with processing emotions such as fear, anxiety, and pleasure.

The experiences we encounter during childhood can have a profound impact on the amygdala. In moments of extreme fear or perceived threat, the brain activates a freeze response as a defence mechanism. Have you ever felt extremely fearful or anxious?

Traumatic events or unsettling situations have the power to trigger fear responses, even without our conscious awareness, as they become lodged in the depths of our subconscious minds. As the amygdala is geared up for survival, the more hyperactive it is, the more signs of post-traumatic stress disorder (PTSD) is present. They linger, influencing our thoughts, emotions, and behaviours long after the initial impact has faded. The subtle remnants of these experiences, hidden away in the recesses of our brains, shape our responses more than we realise.

Prefrontal Cortex

Within a specific region of our brain called the prefrontal cortex lies a remarkable ability to regulate our attention and focus and have rational thinking. This fundamental function allows us to divert our attention from pain and home in on cognitive activities that demand mental engagement, like solving a challenging crossword puzzle. However, it's important to note that this area of our brain is unable to erase memories.

One memory that plays in my mind took place during my early twenties on a dark January evening as I walked alone across a park, a shortcut to my house. My mother's cautionary words echoed in my ears: "I told you not to walk across the park in the dark." Suddenly, a stranger appeared, following me off the bus, and without warning, delivered a forceful blow to my head from behind. I let out a piercing scream as I fell to the ground. Thanks to the cushioning effect of the grass, my fall was softened, saving me from further harm.

Do you find yourself haunted by a similar incident

that replays incessantly in your thoughts? Or is there one that you go out of your way to avoid?

If trauma has taken place for you, a grounding technique may help you to connect the present moment and your physical senses, especially if dealing with flashbacks and hypervigilance.

Exercise

Sight: Look around a room and find specific objects or colours. Pay attention to their details, shapes, or textures.

Sound: Focus on the sounds around you, the humming of appliances, chirping birds, and passing cars. Engage in calming sounds like soft music or sounds of nature.

Touch: Feel the sensation of an object in your hands. Focus on its texture, temperature, weight, and how it feels against your skin.

Smell: You can smell essential oils, scented candles, and fresh flowers. Take deep breaths and focus on specific smells filling the air.

Taste: Slowly savour a piece of food or drink. Pay attention to the texture, taste, and temperature as you chew or sip.

Placebo and Nocebo Effects

By viewing the symptoms you experience from a perspective of safety, you can effectively strengthen the prefrontal cortex. This fascinating part of the brain also plays a significant role in placebo and nocebo responses. The placebo effect generates positive outcomes, especially in pain relief, through the power of our expectations, beliefs, and reward mechanisms. Clinical trials have even demonstrated that administering sugar pills with the belief that they are pain-relieving interventions led some participants to report a reduction in pain.

Nocebo responses, however, are characterised by negative expectations or beliefs regarding pain. Such beliefs can intensify pain even before it occurs, triggering fear responses. For instance, if a person firmly believes that a particular medication will cause side effects, he or she may experience heightened pain due to that negative expectation or belief.

In a publication by Linda Rath for the Arthritis Foundation, a clear connection is established between the brain and the experience of arthritic pain as well as biopsychosocial pain. This understanding has paved the way for a treatment approach known as biopsychosocial pain management, which recognises the significance of an individual's beliefs, expectations, and emotions in his or her overall wellbeing and pain management. By considering these psychological and social factors alongside the biological aspects, a more holistic and comprehensive approach to pain relief and improved quality of life can be achieved. It was apparent in the publication that there was a direct link between the brain and arthritic pain.[8]

I've had a lot of worries in my life, most of which never happened.
– Mark Twain

Hippocampus

The hippocampus plays a vital role in memory formation, learning, and the creation of new neural connections for sustained engagement. It is vital to rewire our brains with positive effects to influence the perception and processing of pain. This specific region of the brain also plays a role in the repetitive thoughts that we experience, along with the tendency to catastrophize and be hypervigilant towards pain based on our memories.

Does the Pixar animated movie *Inside Out* resonate with you? It beautifully portrays the story of an eleven-

8   See: https://www.arthritis.org/health-wellness/healthy-living/
managing-pain/understanding-pain/pain-brain-connection.

year-old girl named Riley, focusing on the emotions inside her mind – such as joy, sadness, anger, and fear – and how they impact her daily life. How frequently have you experienced these different emotions, and how have they influenced your life, particularly in the face of life's challenges? The underlying message in the film lies in understanding and accepting our emotions, acknowledging their presence in our lives, and recognising their importance.

Brain Alteration

The changes occurring in the brain can often result in an exaggerated pain response, triggering false alarm signals that transmit pain signals to the nervous system even when there might not be an actual source of pain. It's akin to a game of Chinese whispers, where the original message gets distorted along the way.

In a thought-provoking article published by *The Guardian* newspaper in May 2023, researchers explored the discovery of specific brain signals associated with chronic pain. Clinical trials were being conducted on deep brain stimulation, a technique involving the delivery of electrical pulses to the brain. While this method has been used successfully in treating Parkinson's disease, addressing chronic pain is far more intricate. The brain regions targeted for each individual may vary considering factors such as their social, environmental, and psychological makeup.[9]

Now, take a moment to transport yourself back to a milestone in your life, perhaps turning twenty-one years old. Reflect on the feelings, thoughts, and emotions you experienced during that time. It's fascinating to realise what you saw, heard, and felt during this milestone, and they would differ significantly from another person's perspective. What remains unknown is the accuracy of the communications or miscommunications occurring between different brain regions. What kind of data are we providing through our thought processes? Is the information we feed our brains precise or distorted?

How Your Nervous System Takes Hold of Your Power

The attempt to escape from pain, is what creates more pain.
– Gabor Mate

It was not until I consciously started paying attention, educating myself, and understanding the true underlying causes of my pain that I made this

---

9   See: https://www.theguardian.com/society/2023/
     may/22/scientists-discover-brain-signals-for-chronic-
     pain#:~:text=Deep%20brain%20stimulation%20sends%20
     electrical,precisely%20which%20signals%20to%20target.

profound connection. It became clear to me that my physical symptoms were linked to the emotions stirring within me.

In a state of panic, I became fixated on seeking quick relief, leading me to explore various avenues such as physiotherapy and numerous visits to chiropractors, hoping for an external fix. While chiropractic treatments provided temporary relief, they didn't offer a long-term solution. To my dismay, I couldn't even perform the exercises recommended as they only amplified the pain I was already experiencing. Interestingly, on some of my appointments, my usual chiropractor wasn't available due to his frequent battles with migraines, forcing him to take a few days off to recover. The irony of the situation struck me.

This experience highlights a crucial point. How can we ever hope to overcome our pain if we remain unaware of its root cause? Unless we dig deep and truly understand the underlying factors contributing to our pain, we will struggle to take effective steps towards finding lasting relief and healing.

Some of the medical professionals advised me to refrain from walking or jogging for six weeks, leaving me feeling even worse and completely powerless. I felt that I had to "be careful." However, during a conversation with Georgie Oldfield, a different perspective emerged. She emphasised the importance of *listening* to our bodies and highlighted that engaging in activity is crucial to healing.

Sensitization

During the experience of chronic pain, it is common for the nervous system to undergo sensitization, whereby the nerves in your body transmit pain signals that are intensified or prolonged. This sensitivity can manifest in two ways: peripheral sensitization, which occurs in the tissues at the site of injury or inflammation, and central sensitization, which takes place within the central nervous system, particularly the spinal cord and brain.

The nervous system plays a crucial role in reacting to perceived threats through the fight, flight, freeze, and fawn response. When the amygdala senses danger, it triggers physiological reactions that prepare the body for action. This response involves the release of adrenaline and cortisol, which increase heart rate, blood pressure, respiration, and other changes that prime the body to either confront the threat or make a quick escape.

Pain Reprocessing Therapy

In an insightful article published by Nathaniel Franklin in the *Washington Post* in October 2021, it was confirmed that the brain has the power to influence pain. The study discussed in the article involved 151 participants who were experiencing persistent back pain. These individuals were divided into three groups:

1. No treatment
2. Given a placebo
3. Given eight one-hour sessions of pain reprocessing therapy (PRT) developed by Alan Gordon, himself a former chronic pain sufferer.

To everyone's astonishment, 66 per cent of the participants who underwent PRT reported either significant or complete relief from pain. Remarkably, these positive results persisted even after a year since the therapy sessions. This finding highlights the extraordinary capacity of the brain to play a critical role in transforming and alleviating chronic pain experiences.[10]

Experiencing a traumatic event – whether physical, emotional, or psychological – can deeply impact your nervous system and overall wellbeing. While you may have had counselling to address the trauma, you may still be experiencing persistent pain unknowingly connected to the traumatic experiences you endured. The underlying thought processes that occur within the brain often go unnoticed. You must

---

10 See: https://www.washingtonpost.com/outlook/2021/10/15/
    chronic-pain-brain-plasticity.

gain understanding and support to help the brain and nervous system feel safe.

The primary role of the nervous system is to protect you from perceived threats and ensure you are safe. However, with chronic pain, the system is stuck in a constant state of false alarm, perceiving a threat even when there is none. Neural circuits in the brain misfire and require consistent rewiring to ease symptoms.

Dysregulated Nervous System

A dysregulated nervous system results from an imbalance between the sympathetic and parasympathetic branches of the autonomic nervous system, leading to heightened pain sensitivity. The brain's interpretation of pain signals amplifies the perception of pain, causing it to feel more intense and overwhelming. This can manifest as a feeling of being on edge, constant anxiety, or difficulty in controlling our emotions.

Power is not given to you. You have to take it.
– Beyoncé

Embracing Pain as a Catalyst for Growth

It's important to understand that the fear of feeling initial pain is natural, especially when reactivating muscles that have been dormant for months, or years. Navigating movement amidst excruciating pain is daunting, and who can blame you if you want to sit back down or just stop? The brain wants you to fear physical suffering as its natural protector is pain. It unfortunately distracts you from dealing with emotional issues we shouldn't avoid.

Regrettably, individuals who have suffered trauma may find themselves resorting to coping mechanisms with serious consequences. These include overdependence on medication or substance abuse, which can even lead to self-harm.

It's Not Your Fault

All too often, a lack of understanding surrounds chronic pain, leaving many in the dark about its underlying causes and complexities. We start to question ourselves, whether we have somehow caused our suffering, or maybe we should have taken a different path. The constant self-blame and criticism become an unwelcome companion in our minds.

Adding to this burden is the social stigma that dismisses chronic pain as, "It's in your head", which only makes you feel worse. The unnecessary pressure society places on us to always appear active and healthy weighs heavily upon those living with chronic pain which adds to a perfectionist mindset. It's only natural to compare ourselves with others and perceive ourselves as failures for being unable to "fix" the pain. This leaves us feeling guilty with agony, fuelling the belief that we are responsible for our suffering.

Invisible Pain

Chronic pain is often invisible which can lead to a lack of understanding or empathy from friends and family. This invalidation can compound doubts about our pain experiences. However, blaming ourselves does not align with the practice of self-compassion. Practising self-compassion is super-important to stopping negative self-talk. Have you ever told yourself how wonderful you are? When was the last time you were kind to yourself? If a friend spoke badly of himself or herself, what would you say to that person that you wouldn't say to yourself? Emotional resilience is something to build upon when the pain is overbearing, especially during flare-ups. If you feel nobody is there to give you a physical hug, give yourself a hug by wrapping your arms around yourself. Hugs are an amazing way to feel warm, loved, and happy. And it can help to reduce pain by increasing the blood flow through the tissues.

Jules Kelly, author of *The Fading Woman*, overcame years of suffering from chronic pain and fibromyalgia. She explained to me that "The fear of staying in the place I was, was far greater than the unknown ahead of me."[11]

Words alone cannot convey the relentless battle we wage daily against an invisible adversary. The pain, hidden from public view, is a constant companion, a silent battle that only we endure. Its intensity is difficult to comprehend for those who haven't experienced it first-hand, and it even suggests to us in a helpful manner that we need an external fix.

It's important to remember that everyone's pain is different, and it's not helpful to compare or dismiss someone else's experience. By prioritising kindness towards ourselves and establishing firm boundaries involving less people-pleasing, we can initiate a transformative shift in how we overcome chronic pain.

Life is 10% what happens to us and 90% how we react to it.
— Charles R Swindoll

---

11 Jules Kelly, Co-Managing Director of Space and Freedom Limited.

Unprocessed Emotions

Dr Sarno's groundbreaking work revealed that chronic pain is often a manifestation of unprocessed emotional tension. He encouraged patients to engage in journaling and mindfulness practices to uncover these hidden conflicts and alleviate pain. It's crucial to remember that chronic pain is not a personal failure. Our brains, conditioned from childhood, generate these sensations to distract us from underlying emotional distress. However, once the pain takes hold, it can become relentless.

The physical changes in our muscles, ligaments, nerves, and tendons, triggered by the brain's distress signal, are what we experience as chronic pain. This invisible condition often baffles medical professionals, who may struggle to identify the cause after extensive testing.

Chronic pain sufferers often feel exhausted, unheard, and dismissed. As the Mayo Clinic aptly points out in its article "What Those with Chronic Conditions Wish Their Friends Knew," even loved ones may fail to grasp the true depth of our struggle.[12]

The shame of feeling inactive, overweight, and physically misaligned when we once were active and

---

12 See: https://mcpress.mayoclinic.org/living-well/what-those-with-chronic-conditions-wish-their-friends-knew/.

capable is a heavy burden to bear. Low self-esteem, feelings of ugliness, and weakness become formidable obstacles. We start to lose our sense of self, concealing our pain deep within.

The greatest challenge lies in confronting our inner turmoil, observing the pain within our minds and bodies, and gradually uncovering a path to healing. It's a minefield of emotions requiring immense mental strength, perseverance, and resilience. We must learn to let go of our self-imposed expectations, embrace the journey of healing, and regain our power.

Exercise

Take a moment to tune into your body and identify where anxiety manifests within you. For me, it often clenches my chest while leaving my mouth dry.

Now, shift your attention to your breath. Observe its rhythm. Is it fast or slow?

Let's explore a simple technique called box breathing, also known as square breathing, which can promote relaxation and reduce stress. Box breathing is counting four in, four hold, four out, four hold, and so on.

Inhale deeply through your nose, counting

to four. Exhale fully through your mouth, also counting to four. Repeat this pattern for four breaths. This will regulate your mind and body. You can do this anywhere as it is effortlessly adaptable to any environment – though remember to keep your eyes open if you're driving.

If a specific trigger exacerbates your anxiety, transport yourself back to a time when you felt calm, perhaps being on holiday or spending time with loved ones. Recall the sights, sounds, sensations, aromas, and flavours that gave you a sense of tranquillity.

Check out my Facebook live with Doc Tovah, chiropractor and presenter of TMS Round Table Broadcast (https://www.youtube.com/watch?v=4v2rC5pvOcQ), who gives hope to others by interviewing medical experts and chronic pain sufferers worldwide.

CHAPTER 3

Believe That the Power
Is in Your Mind

The most common way people give up their
power is by thinking they don't have any.
– Alice Walker

Prior to embarking on mind-body work, it's crucial
to rule out any underlying physical issues that might
be causing your pain. This is essential to alleviate the
brain's fear and doubts, allowing it to engage more
effectively in the healing process. I recall my journey
of scepticism as I grappled with the concept of mind-
body healing. Despite the overwhelming evidence,
my subconscious mind kept whispering fears,
urging me to seek external validation from medical
professionals. My nervous system, still reeling from
the constant pain, yearned for soothing, but my mind
wasn't ready to accept the psychological root of my
suffering.

As a personal injury lawyer, for years I was accustomed to reviewing medical records of individuals ranging from repetitive strain injuries at work to whiplash injuries in road traffic accidents. I couldn't fathom the idea that I, a seemingly stoic individual, could be experiencing similar symptoms. I attributed my pain to the many challenges I had faced in life, subconsciously believing that I was somehow different from these individuals. It took time and introspection for me to acknowledge that I was no different from those I represented or defended. I had to let go of my ego and embrace the profound connection between mind and body, recognising that my pain was not a weakness but an opportunity for healing and becoming powerful once more.

Taking time to do nothing often brings everything into perspective.
– Doe Zantamata

Holistic Approach

Overcoming chronic pain requires a holistic approach that addresses not just the physical symptoms but also the underlying psychological factors that contribute to it. While seeking medical guidance is essential, the path to recovery ultimately lies within us. Staying curious and educating ourselves about the mind-body connection helps us understand the role our thoughts, emotions, and beliefs play in shaping our experiences.

Lisa Feldman Barrett, a neuroscientist and author of *How Emotions Are Made*, emphasises that we have more control over our emotional lives than we may believe. To cope with overwhelm, she suggests the following tools:

1. Take care of your body: Prioritising getting enough sleep and adopting healthier eating habits. Being kind to others also benefits your own wellbeing.
2. Change your environment: Engage in activities that shift your sensory experience, such as going for a walk in nature or moving to a different room. If social media makes you feel uncertain, consider avoiding it.
3. Shift your attention: Practice focusing on your breath, reflecting on something you are grateful for, or recalling awe-inspiring experiences.
4. Change the remembered past: Engage in activities like reading, trying new things, watching uplifting movies, or meeting new people to create positive memories and experiences.[13]

These techniques empower us to have greater control over our emotions.

---

13 See: https://www.mariashriversundaypaper.com/lisa-feldman-barrett-cope-with-distress/.

To further explore this concept, you can ask yourself the following questions:

1. What would make you believe you can become pain-free?
2. What beliefs did you grow up with that you later realised weren't true?
3. What kind of person do you need to be to achieve a pain-free state?
4. What limiting beliefs do you currently hold?

Exercise

Understanding Your Circles of Influence

Stephen Covey, the author of the renowned book *The 7 Habits of Highly Effective People*, introduced the concept of circles of influence to help individuals understand the areas where they can exert control and those where they cannot.

Visualising Your Circles

Identify your concerns: Begin by creating a list of all the issues or concerns that occupy your mind.

Define your circle of concern: Draw a large circle on a piece of paper, and label it as your "circle of concern". Within this circle, list all the items from your concern list.

Discern your circle of influence: Inside your circle of concern, draw a smaller circle representing your "circle of influence". Carefully evaluate each concern in your larger circle, and move those you can actively influence into the smaller circle. What do you observe?

Establish your circle of control: If desired, draw a third, even smaller circle within your circle of influence, representing your "circle of control". These are the concerns that fall directly under your control. Transfer the relevant items from your circle of influence into this innermost circle.

The Power of Awareness

By visually representing your circles of influence and control, you gain a clearer understanding of the areas where you can effectively channel your energy and focus. This exercise helps you prioritise your efforts, recognise where change is possible, and accept what lies beyond your control.

The Power of Community

The soothing voice of Georgie Oldfield told me that this alien pain I was experiencing was normal and unfortunately, self-inflicted in my unconscious brain. I finally felt safe, and my hope was reignited that I would be able to recover. I was finally being *seen* and *heard*.

Unbeknownst to you, the power to heal lies within you. No one has told you that you hold the key to your recovery, that you can manage and alleviate your pain without relying solely on traditional medicine. The sad reality is that many people with chronic pain are left to go on this journey alone, without the guidance and support they desperately need.

Peer-to-peer support is a beacon of hope in this isolating journey. Connecting with others who share the same struggles can provide invaluable validation, understanding, and encouragement. You'll find strength in their stories, resilience in their perseverance, and a sense of belonging in their shared experiences.

Whether through one-on-one interactions, group sessions, or online communities, seeking peer-to-peer support will boost your confidence and rekindle hope. However, it's crucial to find a supportive environment that aligns with your healing goals. Avoid groups that focus solely on symptom sharing and anxiety. Instead, seek groups that promote positive coping mechanisms, foster a supportive atmosphere, and encourage individual growth.

Awareness and unwavering belief are the cornerstones of your recovery journey. Understanding the root cause of your pain – whether physical, emotional, or a combination of both – will empower you to take the necessary steps towards healing.

Reaching Out

Asking for support is an act of courage and self-compassion, a necessary step in reclaiming your wellbeing. Asking for help was a challenge for me after dealing with emotional issues alone as a child and fearing rejection. If you're sensitive, you may worry about judgement when opening up. However, local communities are being formed by medical professionals, charities, and individuals with lived experiences who have overcome or significantly reduced their chronic pain. These communities offer support, regular communication, and even coaching, often involving activities like yoga or walking.

If you've always depended on self-reliance, reaching out for support may feel daunting. You may question the effectiveness of support, wondering if others can truly understand your unique experience. But remember, you deserve to be seen, heard, understood, and supported. Trust in your ability to find the support that resonates with your needs, and allow the collective wisdom and compassion of others guide you on your path to healing.

Findings from a 2016 study published in *Research Gate* highlight the benefits of interpersonal connections for individuals with chronic pain. One war veteran shared, "Just taking time out, discussing, relaxing, and forgetting things for an hour or so was most beneficial." Another veteran, who had been socially isolated, found solace in simply talking to someone, "not necessarily about pain, just talking to another person." The common thread was that *communication* alleviated stress, underscoring the importance of peer support. While self-management is possible, most of us benefit from a helping hand and nurturing support.[14]

Lack of Signposting

The lack of signposting following medical consultations and scans can leave individuals feeling lost. When there's no open discussion about childhood experiences or stress triggers, individuals tend to remain silent, placing excessive hope in the

14 See: https://www.researchgate.net/publication/305213781_
How_Do_Patients_with_Chronic_Pain_Benefit_from_a_Peer-
Supported_Pain_Self-Management_Intervention_A_Qualitative_
Investigation.

medical profession. This can lead to disappointment when traditional treatments fail.

To bridge this gap, it's crucial to actively seek support. Online and in-person communities provide safe spaces to connect with others who understand your struggles. Their stories, resilience, and shared experiences can offer hope, validation, and encouragement to keep moving forward.

Your wound is probably not your fault, but your healing is your responsibility.
– Denzel Washington

Amidst the isolation of lockdown, I found solace in Georgie Oldfield's private Facebook group, a global sanctuary for chronic pain sufferers. Vulnerable and disheartened, I sought support from fellow chronic pain warriors, their shared experiences a beacon of hope in my darkest hour.

Patricia Stanghellini, a compassionate friend and coach from Italy, extended a helping hand, guiding me through the labyrinth of low self-esteem. Chronic pain had eroded my confidence, leaving me yearning to feel beautiful and vibrant once again. Seeking long-term support, I reached out to Jeannie Kulwin, a coach with a profound understanding of chronic pain. Howard Schubiner and Steven Ozanich, along with my family, were also within my support network who helped reassure me that I would be able to recover.

Within these supportive circles, magic unfolded. I felt truly heard and championed, empowered to embrace a mindset that promoted healing. Openly sharing my pain without shame, I drew strength from the stories of others who had overcome similar struggles.

Surrounded by this inner circle of understanding, I was no longer alone in my journey. Their unwavering belief in my resilience instilled hope, fuelling my determination to reclaim my life. I began to grasp that my pain was not a physical defect but a neurological battle, a challenge to be conquered through the power of my mind.

With newfound determination, I resolved to emerge victorious in this internal war of me versus me. I would not let chronic pain define me. Instead, I would reclaim my strength, my confidence, and my life.

The harder the struggle, the more glorious the triumph.
– Unknown

The Path to Healing: Embracing Belief and Taking Action

Each individual's perception of pain is unique and shaped by life experiences, beliefs, and environments. Our subconscious minds often have limiting beliefs like, *I'm not good enough,* or, *I'm a misfit* which can hinder our healing journeys.

Imagine reaching your destination, a life free from chronic pain, a life you never dared to dream possible. This transformation is within reach when we address the limiting beliefs, emotions, and thoughts that hold us back.

Beyond Peer Support

Sharing your experiences can be cathartic, but it's crucial to balance this with action. Relying solely on these groups for validation without taking concrete steps towards healing can be counterproductive.

While social media can provide inspiration, tips, and a sense of community, it's important to use it responsibly. Excessive exposure to social media can exacerbate anxiety and fear, hindering your progress.

Breaking Free from Digital Obsession

In today's hyperconnected world, our brains are often bombarded with digital stimuli. Constant notifications, endless scrolling, and the pressure to stay connected can take a toll on our mental health.

Johann Hari, in his book *Stolen Focus*, highlights the

negative impact of our digital habits. Setting limits on social media consumption, turning off notifications, and engaging in non-digital activities can help detox our brains and promote healing from our pain.

Faith includes noticing the mess, emptiness, and discomfort, and letting it be there until some light returns.
– Anne Lamott

Here's a link to my podcast with Eddy Lindenstein: https://audioboom.com/posts/8288963-272-narinder-sheena-former-lawyer-gives-her-amazing-recovery-rundown.

Exercise

Draw three circles, one inside the other. In the inner circle, write down all the people you can call anytime without hesitation – your close friends, family members, or anyone who provides you with unwavering support. In the second circle, list the individuals you can reach out to, although you may feel that they may not always be available. Finally, in the outer circle, write down other valuable connections in your support network, such as work colleagues, teachers, doctors, or dentists.

This exercise proves helpful if you find it challenging to open up and share with others, or if you believe you lack someone to confide in. Even if you feel that your inner circle is empty, consider exploring the individuals in the other circles. Assess whether you can *trust* them and feel comfortable sharing your feelings and seeking support from them.

I initially found this difficult as I believed I could handle everything on my own. However, this proved detrimental as I struggled in silence. It was only after I mustered the courage to reach out that I found the support I needed to embark on my healing journey.

Coaching: Unleashing Your Power to Conquer Chronic Pain

Surround yourself with people that reflect who you want to be and how you want to feel. Energy is contagious.
– Rachel Wolchin

The role of a coach is to motivate and champion you, helping you unlock your potential to achieve your goals or overcome resistance. Sometimes we may not be aware of our resistance. We all have areas where we strive for perfection, but it's important to focus our energy and commitment.

Goal Setting and Self-Reflection

Coaching facilitates deep introspection, allowing you to confront your thoughts and emotions with clarity. Open-ended questions guide you towards positive actions and forward-thinking solutions.

Active Listening: A Healing Touch

Many chronic pain sufferers have never experienced true, empathetic listening. Nancy Kline's book *A Time to Think* emphasises the importance of deep, attentive listening. A client of mine who had suffered chronic back pain for years expressed her gratitude for being heard as she usually avoided opening up to such an extent. Truly listening to a sufferer's experiences, frustrations, and emotions provides immense support for his or her healing journey.

Building Trust and Connection

Chronic pain sufferers often face judgement from others – including family, friends, and even medical professionals – due to the invisibility of their pain. Building trust and connection is crucial for their healing. As a coach, it's essential to hold space for emotional release, allowing clients to process their thoughts and feel validated.

Noticing Body Language and Energy

A coach identifies energy shifts in clients, such as slumped shoulders, fidgeting, or a mismatch between words and physiology. Facial expressions or lack of eye contact also convey unspoken emotions. Through deep listening, coaches can raise awareness, helping clients overcome energy blocks, low confidence, and lethargy.

Accountability and Self-Compassion

Accountability supports your healing journey. Tracking progress, setting achievable goals, and using tools like vision boards or Mind Maps can enhance motivation. Remember, self-compassion and kindness towards yourself are essential elements of healing.

Self-Awareness: Unlocking the Inner Healer

A coach helps you to identify and address negative thought patterns, enabling you to overcome stress, anxiety, overwhelm, and fear. Through open questioning, coaching helps you gain self-awareness, explore your current reality, and identify emotional triggers.

Nurtured Support for Your Healing Journey

Coaching provides a confidential, non-judgemental space for growth and transformation. Through open-ended questions, a coach partners with you towards clarity, empowering you to take ownership of your healing journey.

A study published by the Integrative Pain Science Institute examined the impact of health coaching on individuals with fibromyalgia. The researchers found that those who received health coaching experienced significant improvements in pain and reduced the usage of healthcare services.[15]

Rather than being your thoughts and emotions be the awareness behind them.
– Eckhart Tolle

---

15   See: https://integrativepainscienceinstitute.com/health-coaching-and-fibromyalgia/.

Embrace the Pain:
Confront Doubt with Resilience

I stumbled upon an article suggesting that instructing my brain to release muscular tension could alleviate pain. Embracing this idea, I embarked on a journey to rewire my neural pathways, encouraging blood flow to my affected leg. This involved soothing my mind while in motion, gradually improving my condition. I continuously reminded myself, "I am fine", cultivating a belief system that countered the pain's persistence. Any doubt would rekindle the pain, fuelling my determination to persevere and reinforcing my conviction in self-healing.

The Curable mobile app introduced me to a transformative exercise that revolutionised my ability to relax my nervous system, dispelling fear and fostering tranquillity. It involved conversing with my brain aloud, requesting it to intensify the pain through a countdown. Recalling how sitting on my soft sofa would exacerbate the burning sensation in my groin and buttocks, I would boldly challenge, "Come on, give me more pain." After a minute, the pain would plateau, followed by a reduction, igniting a spark of hope and propelling me forward each day. This exercise reinforced the notion that my brain was the pain generator.

Shifting from physical to psychological thinking, a

daunting task for the unconscious mind, is a crucial step towards pain liberation. Identifying the triggers that precede or coincide with chronic pain flare-ups is essential. Once the unconscious mind accepts the psychological nature of pain and feels safe, the pain diminishes. However, convincing the unconscious mind takes time and patience, as it's not as readily persuaded as the conscious mind. Remember, the *pain is not dangerous*. It's a manifestation of the mind's interpretation of physical sensations.

> *The only way out is through.*
> – Robert Frost

Myths of Chronic Pain

In an article by Zach Walston published in *Physical Therapy (PT) Solutions* in August 2020, some common myths and misconceptions surrounding chronic pain were highlighted:

Myth: The pain is in your head. Chronic pain is a complex condition that causes real pain to sufferers. While the brain's neural pathways may contribute to the pain, it is influenced by physical, emotional, and neurological factors, not just psychological aspects.

Myth: Hurt always means harm. Chronic pain persists because the brain generates pain signals, leading to a heightened nervous system and a false sense of danger in the body. This fear-pain cycle can result in ongoing pain, but it doesn't mean there is physical harm.

Myth: Pain means you need to stop all physical activities. Gradual physical activity, at your own pace, is one of the best ways to strengthen your body, improve mobility, and enhance overall well being. Being inactive, or not doing any activity can be counterproductive and can lead to depression.

Myth: Pain medication is the only effective treatment for chronic pain. While pain medication can provide temporary relief, relying solely on medication can lead to dependence without addressing the root cause. Exploring other holistic approaches and pain management techniques can be more beneficial in managing chronic pain.

Myth: People with chronic pain are just seeking attention or exaggerating their symptoms. Many individuals with chronic pain have a history of emotional problems, which can contribute to the persistence of their pain. It is essential to approach pain sufferers with empathy and understanding rather than assuming their intentions or downplaying their symptoms.

Myth: Chronic pain is something you need to live with or manage permanently. It is possible to significantly reduce pain and even live pain-free, like me, even if you have experienced long term pain. Therefore, through proper understanding of your own self care, you are able to eliminate it.[16]

In the *American Journal of Neuroradiology* (AJNR), Waleed Brinjikji published a study in 2015 ("Systematic Literature Review of Imaging Features of Spinal Degeneration in Asymptomatic Populations") that revealed degenerative spinal changes in the MRI scans of healthy forty-year-olds. This finding indicates that such changes are common and not necessarily alarming.[17]

---

16 See: https://ptsolutions.com/blog/addressing-pain-myths/.
17 See: https://www.ncbi.nlm.nih.gov/pmc/articles/PMC4464797/.

Questions to Ask Yourself

To further explore your personal pain experience, here are some questions to consider:

1. On a scale of 1–10, how motivated are you to become pain-free? Be honest with yourself about your level of commitment.
2. What would it mean to you if you recovered from the pain that has been affecting you?
3. How frequently do you engage in daily activities – rarely, once or twice a week, or often?
4. How often do you experience anxiety?
5. Does your pain increase when you have increased anxiety?

Exercise

1. Make a list of all of your strengths, and focus on them every day.
2. Track occurrences of migraines, back pain, or other discomforts, and try to identify any triggers consciously.
3. Write down any self-blaming thoughts that arise, and reflect on their truthfulness.
4. Practice gratitude by writing down or verbally expressing three things you are grateful for every day.

CHAPTER 4

Reclaim Your Power: Shifting Your Mindset

The greatest glory in living lies not in never falling, but in rising every time we fall.
– Nelson Mandela

The negative and overwhelming emotions we experience can drain our energy, especially when we are already dealing with persistent physical pain. It's important to treat chronic pain with compassion, allowing it to linger as long as necessary because it serves a purpose and carries a significant message that we are required to listen to our bodies. The pain of movement may have sapped all your energy, leaving you exhausted. It's understandable that you may feel hesitant to join in on activities and are unable to walk, sit, or stand comfortably. Who wants to be the one to dampen the mood during a pleasant evening of dinner, drinks, a coffee catch-up, or an exercise class with friends? You may have grown accustomed to this reality as it has been a long time since you experienced the joy of laughter.

The Use of Words That Can Cause Fear

I distinctly remember the moments when medical professionals warned me to be vigilant for signs of cauda equina syndrome (loss of bladder control). This caused me to constantly monitor myself as they advised seeking urgent medical attention if I couldn't feel my bladder movements. I used to have frequent trips to the bathroom, thinking it was due to having children, only to discover that it was a manifestation of my anxiety. My mind immediately jumped to the worst-case scenario: fearing the need for emergency surgery. Your body is stronger and more resilient than you think. Stopping yourself from moving a lot actually keeps you in pain much longer.

It is natural for the emotional impact of chronic pain to consume our thoughts, but it is within our power to reshape that narrative. By embracing the belief that there is a different path and committing to making a change, we can regain control over our lives. It's important to acknowledge that *anxiety* plays a significant role in fuelling chronic pain. Stress is the feeling of being overwhelmed or under intense pressure. It can manifest anxiety within us, often arising from various sources and directions.

Recognising this connection between stress, anxiety, and chronic pain is a crucial step towards finding relief and implementing effective coping mechanisms.

It's understandable that you might feel deeply impacted by depression when you're constantly experiencing pain. It's no laughing matter to be confined to the floor, twisting and turning your body in search of even the slightest bit of relief. Perhaps sleep becomes an escape, an attempt to wish away the pain. It can feel like torture, a relentless blow to the head that leaves you wanting to get out of bed, but the overwhelming thought of being unable to function holds you back. You may find yourself berating and devaluing yourself, feeling upset, and continuously reliving painful memories that loop in your mind.

Imbalance in the Brain

The imbalance of chemicals (neurotransmitters) in the brain, such as serotonin (which carries messages between nerve cells) and norepinephrine (which helps transmit nerve signals), plays a crucial role in regulating both your mood and your perception of pain.

Interestingly, anger can amplify your pain within the central nervous system, triggering physiological responses like muscle tension, increased heart rate, and hormonal fluctuations. Anger often arises as a defence mechanism, and it can evoke a sense of helplessness. It's important to note that anger and frustration can become stored within the body, held tightly in a secret place, like a hidden safe, away from prying eyes. This stored tension adds a layer to the complex interplay between emotions and physical pain.

Energy

Our energy levels are intricately linked to the way we think, manage our emotions, and the emotional influences we receive from others. This question arises: Are we in control of our emotions, capable of managing our stress responses, emotional expressions, and self-awareness, or do they control us? I noticed how watching the news affected my unconscious brain. I'm curious, do you watch the news, and if so, does it impact your overall wellbeing?

There's a saying, "Your energy introduces you before you speak," which holds true in various social contexts, be it at work or in personal interactions. Take a moment and reflect on your conversations. Do you feel drained afterwards? Are you receiving the support you need, or does someone monopolise your time? It is essential to surround yourself with individuals who will uplift you, listen to, and care for you. Lean on these people as you embark on your journey towards recovery.

Reassurance

When I reached out to Georgie Oldfield, I was desperately seeking answers for my stress-induced symptoms. I was very vulnerable, overwhelmed, and panicked. I explained to her that I could barely walk or stand. Instead of the expected response of sympathy, she pleasantly surprised me with words of hope and guidance. She advised me to reframe my thoughts and acknowledge my fears, assuring me that I would recover.

These words of hope were a stark contrast to the experiences I had with other medical professionals. Instead of instilling confidence in my recovery, they only fuelled my fears, emphasising what I couldn't do and making me feel restricted in my ability. Georgie's belief in my potential felt refreshing and made me question my lack of trust in myself.

The brain is an incredibly intelligent organ and cleverly redirects manifestations of stress to different areas of the body once one area has recovered. I used to suffer from migraines, attributing them solely to lack of sleep due to having a baby and leading a hectic lifestyle. While these factors likely contributed, the true underlying cause was the stress I was experiencing.

Growth

In her book *Mindset*, Carol Dweck emphasises the concept of the growth mindset, which highlights that individuals who adopt this mindset believe that intelligence and abilities can be cultivated through effort, practice, and learning from mistakes. They embrace challenges, view failures as opportunities for growth, and persevere in the face of setbacks. This mindset fosters greater motivation, resilience, and achievement. Additionally, it encourages the acceptance of failure as a natural part of the learning process and emphasises the importance of consistent effort, even in the face of challenges.

Instead of saying to ourselves, "I can't walk that far," imagine if we rephrased it as "I can't walk that far yet." This simple shift in perspective acknowledges that growth and progress are possible with time and consistent effort. I never believed I could run a half-marathon, which is approximately just over thirteen miles. However, by consistently challenging myself and practising running a few times a week, I was able to accomplish this feat. Forming a new habit typically takes around sixty-six days. This can vary for each individual, so being persistent and staying consistent are crucial on the journey towards growth and improvement. Accepting the presence of pain is a crucial step on your path to recovery.

The Boulder Back Pain study, published in *JAMA Psychiatry* in September 2021 by Alan Gordon, LCSW, author of *The Way Out*, and Dr Howard Schubiner, a mind-body specialist and author of *Unlearn Your Pain*, offers valuable insights. The study revealed that 66 per cent of chronic back patients experienced either freedom from pain or a significant reduction in pain after just four weeks.[18]

What makes this study particularly intriguing is the finding that the misfiring neural pathways in chronic pain patients activated different regions of the brain than those associated with acute incidents. This emphasises the role of the brain in generating hypersensitivity and hypervigilance. It's important to remember that the amplification of the pain sensations originates within the brain itself.

Neuroplasticity: Rewiring Your Brain

Neuroplasticity refers to the brain's ability to rewire and create new neural pathways. According to a March 2018 article by neuroscientist Tara Swart, "The

---

18 See: https://jamanetwork.com/journals/jamapsychiatry/ fullarticle/2784694.

FROM PAIN TO POWER

4 Underlying Principles of Changing Your Brain",
it is still possible to retrain your brain and develop
new habits even in later stages of life. However, this
requires consistent repetition of new habits to bring
about lasting change.[19]

Lee Vaughan, a fellow chronic pain warrior and
founder of an online community Partnering Pain,
explained to me, "For me, it's protecting the 10 year
old me though, because he is the one that hurts me
today. Reflecting on life I realised that acute trauma
exists in my past and it was only when I began to
learn about pain that I made the connection between
trauma and pain. That was the lightbulb moment I
had been seeking for over 30 years. Then I began to
heal. There is nothing biologically wrong with me.
The pain is 100% neuroplastic."[20]

Now that you have become aware of your negative
and unhelpful thought patterns, it is important to
challenge those beliefs. Begin by questioning the
evidence that supports your thoughts or feelings.
Are they true or based on assumptions? Assess
whether these assumptions are helpful or harmful to
you. Reframing your thoughts is a crucial step. Start
searching for evidence that can replace negative

---

19 See: https://www.forbes.com/sites/taraswart/2018/03/27/the-
    4-underlying-principles-to-changing-your-brain/.
20 Lee Vaughan, Founder of an online Facebook community called
    Partnering Pain.

thought patterns with positive ones. Writing positive affirmations or statements and consciously embedding them in your mind every day can be beneficial, even if you don't initially believe them. Remember that your unconscious brain may resist this change at first. But with persistence and consistency every day, a shift will occur in rewiring your brain.

Positive Self-Talk

In *Psychology Today* (January 2020), Dr Joe Tatta, a physical therapist and ACT trainer, explained that negative thoughts about pain, such as magnifying the power of the pain, constantly ruminating about it, and helplessness, where you feel you cannot recover from chronic pain, are all embedded in the brain. It is by acceptance and commitment therapy (ACT) that three things can help ease the negative pain processing.

1. Cognitive defusion is where you accept some of the pain without judgement and work on your values, and whether you are aligned with those values to know who you are. This will be discussed later.
2. Positive self-talk such as, "I am gaining knowledge to change my pain."
3. Moving with your mind – a simple exercise of noticing the thought that comes to your mind

when moving, naming that negative thought, and then continuing to do the activity.[21]

The following are some of the statements from the SIRPA recovery programme and Dr Sarno's daily reminders I used as part of my healing journey:

• The pain is emotionally induced and not due to a structural abnormality.
• If I have a real injury, why does the pain come and go for no apparent reason?
• I will not be concerned or intimidated by the pain.
• I will shift my attention from the pain to the underlying emotional issues.
• I intend to be in control, not my unconscious mind.

Your beliefs become your thoughts. Your thoughts become your words. Your words become your actions. Your actions become your habits. Your habits become your values. Your values become your destiny.
– Mahatma Gandhi

---

21 See: https://www.psychologytoday.com/us/
blog/psychologically-informed-approaches-pain-
management/202001/3-ways-overcome-negative-thoughts.

Positive Affirmations

Positive affirmations are statements that will help your mindset in a positive manner. It will help shift your thoughts and beliefs towards an optimistic and empowering perspective. When the affirmations, or statements, are repeated regularly, the unconscious mind is getting rewired, boosting your self-confidence.

Here are examples of positive affirmations I used:

- I am healing, and I feel alive.
- I am becoming active, and I feel wonderful.
- I am letting obsessive thoughts go, and I feel calm.
- I am not afraid of the discomfort, and I feel safe.

Acknowledge Your Fears

I would also acknowledge and write down any fears I had before I carried out an activity to help calm my nervous system and reduce my anxiety every time I could feel the sensations in my body. Mine would usually be the chest tightening and fast breathing.

> *Fear* – walking outside of my house. "What's the worst that can happen?" I would ask myself. I will need to stop and sit down, and people will see me. If that happened, then I would

take a break. And if they saw me, I could tell them that I have discomfort in my legs or back. Another way of looking at it is that it's okay to stop and take a break. And if neighbours do see me, it's okay.

Fear – standing longer than five minutes as I feel my muscles will tighten on one side, especially when standing to wash dishes or cooking food. The worst that can happen is that I sit down. It's okay to take a break and then start again.

Fear – waking up in discomfort every morning. The pain is always there. How about I just focus on getting up and visualising that there is no pain in the area I am focusing on – for example, my lower back – and tell myself that I'm feeling wonderful?

Fear – going swimming. I would ask myself, "Will I be able to swim?" I can swim in the knowledge that the water supports my body, and it will help my body loosen my muscles and relax. I can also slowly get changed in the changing room by sitting down and taking my time.

My fears didn't come true. Although there was discomfort walking and going swimming, I still went into the swimming pool and swam. That made me feel confident and grateful that I could do it. It felt amazing just to do that. Afterwards, I felt the uncomfortable sensation, but I put that down to exercising my tight tendons and muscles and the belief that if I carry this on consistently, my body will become strengthened.

Whether you think you can or think you can't, you're right.
– Henry Ford

Howard Schubiner

I reached out to Howard Schubiner via email regarding my chronic pain and where I was feeling it in my body. I told him how I started going swimming, hobbling there, although the water was my biggest soother. I described the pain as a 9 out of 10 when I stood, whereas sitting would reduce it to 2 out of 10. I explained how at the gym I started cycling with a higher resistance, and the pain in my right leg would reduce; I figured the blood flow was circulating. However, the pain in my lower back was still persistent. At this stage, I was unable to walk on the treadmill much as the muscles on my right side used to tighten up.

I expressed how scared I was and asked whether I had "damaged" my leg or hip area. *I cannot hurt myself* did not spring to my mind. Due to my mobility restrictions, I felt depressed, and I focused on the pain. I definitely was not enjoying my life. I told Dr Schubiner it had been two months like this and that I was usually a positive person, but this was really getting me down.

To my pleasant surprise, I received a very quick reply. He confirmed to me that the pain I was experiencing was due to the neural circuits in my brain moving from place to place, which can come and go away. The best advice he gave me was to soothe my nervous system. So, remember if the pain moves around the body, it is a mind-body connection.

Exercise

Write three to five positive affirmations you can see daily, attaching a feeling to the affirmation like mine.

What positive statements can you write to embed in your mind?

Negative Thought Patterns from Trauma

*Trauma isn't what happens to us, it's what
happens inside us.*
– Gabor Mate

Mental health disorders from childhood trauma can lead to anxiety, depression, and PTSD. Trauma can cause changes in the brain's structure and function, which distorts how pain is perceived and persists, as we have seen, within the amygdala. It is naturally associated with individuals who have been in war or natural disasters, just to name a few, where we know how distressing these events can be on someone. These are referred to as "big 'T' trauma" when there are life-threatening events. I was speaking to someone who used to be in the army, and he told me that his friend took a bullet for him, and he suffered PTSD after that. Imagine the knock-on effect. In relation to road traffic accidents, which I would see quite often in my work as a personal injury lawyer, a person could get flashbacks after an accident or avoid certain roads. These events would recur. Witnessing a car crash or a fatality is also prevalent in PTSD.

Trauma can be a situation you are experiencing right now, for example, the fear of losing your job or financial worries, which is referred to as "Little 't' trauma." Although these references may appear small to one individual, they may seem like a huge trauma to another.

Trauma involves so much. Initially, it can be a physical incident such as a road traffic accident, where you can see blood or wounds. However, emotional trauma, as we have seen with ACEs, can happen anytime in our lives. It can be as simple as somebody saying something to us as a child that caused us feelings of shame. We can then be triggered if a similar scenario arises. Remember that children are not able to emotionally regulate or understand how their traumas have impacted them until they are old enough to start becoming aware.

How Trauma Works

The three types of trauma can be any or all of the following:

1. Acute, involving a single incident.
2. Chronic, which is repeated and prolonged, and may involve physical or sexual abuse or war.
3. Complex, where you are exposed to a variety of traumatic events. For example, a child is suffering in poverty and also being physically and sexually abused.

Coaches who deal with clients suffering chronic pain are likely to face many challenges where trauma is involved, especially when a client feels stuck or unable to make changes to help themselves. Deep listening and empathising with what happened in the sufferers' lives, and allowing them to sit with their thoughts and emotions is a great way to process what's coming up for them. And as a coach, acknowledge what they bring and support their self-awareness.

Anita Guru, a coach and public speaker, shared her journey through this wonderful poem:

It's Not Me

It's my illness you see
It takes me to places where I stand alone in the woods
It's dark and gloomy with dead leaves
I don't see a way out of the woods
I try to break through by smashing trees
But they stand so strong and overpowering me
It's the roots deeply embedded
Not allowing such a change
It's too ingrained
I stand exhausted after 8 months of trying
I've shed buckets of tears so I can keep fighting
The pain and hopelessness keeps reigniting
It's my despair as I look at my life
All I wanted to do was be a mother and a wife
My life tale tells a sad story
Barely any moments of glory
Yet I keep looking for a way out
Before my tears cause a drought

The journey continues with no known destination
I need to find it fast, before I end up cremated

(25 July 2018 @poetryknowsnoname [Instagram].)

Catastrophizing

These conditions then heighten the perception of pain through catastrophizing (exaggerated negative thoughts and emotions about pain) and interfere with the ability to cope with the pain effectively. The fear-pain cycle keeps chronic pain alive through the unconscious mind.

According to an article published in the National Library of Medicine (NLM), "The Influence of Adverse Childhood Experiences in Pain Management: Mechanisms, Processes and Trauma-informed Care" in June 2022, 84 per cent of adults who were suffering from chronic pain reported at least one ACE, such as fibromyalgia. The findings were not only of increased physical pain but that these individuals were vulnerable in their psychological wellbeing. Those with anxiety and chronic pain had an increased risk of suicide attempts.[22]

---

22 See: https://www.ncbi.nlm.nih.gov/pmc/articles/PMC9226323/.

The greatest weapon against stress is our ability
to choose one thought over another.
– William James

Chronic pain is not one size fits all. It's complex, and every individual will experience different levels of stress or ACE. Some people don't realise they have experienced ACE, and they will feel stressed. Understanding and listening to a patient as a whole person with empathy, compassion, deep listening, and finding out how he or she is coping is an important place to start in determining whether medication, surgery, or injection is necessary in the first place.

How the Nervous System Plays Its Part

Will this individual benefit from such medication when his or her nervous system is already heightened, and no amount of medication has previously made the person feel better? The steps for a patient to take accountability are huge as it requires a lot of commitment. However, partnering with the patient every step of the way can create a transformational change. When I have clients who are deeply distressed, I notice their physiology, their quickened pace of breath when they are talking, and their facial expressions. These signs tell us that something is happening with these patients that first needs calming down. Maybe some deep breathing or a somatic approach to allow their nervous systems and minds to relax. Nurtured steps to reducing a patient's chronic pain are crucial.

David Hanscom MD, a former orthopaedic surgeon and author of *Back in Control*, explained that humans cannot tolerate emotional pain, which is why dealing with physical pain may help mask that nicely. Chronic pain is memorised through what we know, our beliefs. That is, what we have been told or seen. For example, people who have amputations sometimes feel pain at the site of their amputated limb. The only reasonable explanation is that the pain is generated in the brain.

Do You Have the Willpower to Recover?

The fear of being vulnerable and allowing our egos to shield us from pain is always going to be a stumbling block. If we can acknowledge that we may require help, this will be a huge step in the healing process. It's not a sign of weakness. Rather, it is a step to show that you want to recover.

In an article by David K Harris, MD, "Why an MRI May Not Reveal the True Source of Your Pain" (August 2020), he states that pain cannot be detected through MRI, CT scans, or X-rays. The ageing process will inevitably show imaging of various bones or tissues degenerating over the years, but they are not indicative of damage to the body. It is so important to accept that the pain you feel is not what imaging is going to confirm. It appears logical, as it did with mine, where it showed a severe centralised disc prolapse, and surgery was recommended. If you saw an MRI of someone who was not experiencing any pain, and the imaging was the same as yours, would you then believe that the source of your pain is through your brain?[23]

The fear of waiting for medical appointments can prolong the pain you feel. It only allows the catastrophizing of old or new pain sensations to continue, thinking that the next appointment will give you an answer or solution. Unfortunately, catastrophizing about what you might be told during a long wait for an appointment can in itself cause pain to persist. How long it will take to recover is often dependent on how much belief we have that the pain we are experiencing is psychological.

Reprogramming our unconscious minds so that we are able to recover from our physical pain depends on how much belief is attached to the idea that it has a psychological cause. This can take time to reinforce and embed into our brains.

---

23 See: https://charmaustin.com/why-an-mri-may-not-reveal-the-true-source-of-pain/.

Each individual's perception of pain is going to be different as each of us has lived life through a different lens. Your personality type will differ from that of the next individual and the way you see things. Pain is no different. Nobody knows its *root cause* until we start doing the inner work.

Belief

Believing that you can recover by yourself with nurtured support, without allowing self-doubt, self-sabotage, or resistance to creep in, is necessary. Your thoughts – you know, those gremlins (inner critics) that are working hard to keep you safe – your beliefs, emotions, and other people may steer you away from trusting yourself. Your mind or others will keep insisting again and again that you need a medical professional to fix you. It may have worked for them or for someone they knew who had it, but this one is all on you. When you understand how your brain works and how your nervous system is dealing with the pain, it may give you the motivation to believe you are going to heal.

Pain is temporary. It may last a minute, or an hour, or a day, or a year, but eventually, it will subside and something else will take its place. If I quit, however, it lasts forever.
– Eric Thomas

An article in the *Guardian* by Liam Mannix, "Is Good Posture Overrated? Back to First Principles on Back Pain" (August 2023), highlighted that there was no conclusive evidence that bad posture causes back pain. Our spines are curved, and nobody walks around in a military pose unless you are in the military. Being correctly signposted is important as is having patience. Waiting around for further medical appointments or imaging just fuels the pain.[24]

How resilient are you to believe that if you do the hard work, motivate yourself in knowing 100 per cent you will recover at your own pace, and without comparing others' recovery stories to your own that you will get there? I kept obsessing about how quickly someone recovered from a similar physical problem. However, that was not my journey and would have disheartened me in my progress.

How empowered will you feel when you reach your destination and then discover a transformed life, one you could not even dream was possible, because the limiting beliefs, emotions, and thoughts were replaying into your unconscious? Whatever way you feel is the best approach to help you, just know that you cannot hurt yourself, and you definitely can get there.

---

24 See: https://www.theguardian.com/science/2023/aug/06/good-posture-back-pain-how-to-avoid.

Exercise

How often do you watch the news before going to bed?

Do you feel okay afterwards, or does it affect your thoughts before going to sleep?

What can you do differently to help you have good thoughts?

What bedtime routine helps you to relax before sleeping?

Write down a list of what you can do and what you would like to do. For example, "I would like to walk for thirty minutes by … " and give yourself a deadline.

Given that you have now started shifting your mindset, you can take ownership and accountability to move forward in making your recovery possible.

CHAPTER 5

Take Back Control of Your Power

If you can feel inspired and motivated, and know why you have to do it, digging deeper than you ever have before is likely to help you on your healing journey. The resistance in your brain is going to make it difficult for you, but working on your mindset one step at a time is likely to help overcome those obstacles.

Somatics is a pathway to awakening the body and accessing the deep well of inner peace within.
– Eckhart Tolle

I came across somatic tracking while listening to a podcast, Alan Gordon, LCSW, author of *The Way Out*, who was trying it with someone suffering from back pain. I was intrigued about how this would work on me and the pain I was feeling. Somatic tracking encourages you to observe your pain without judgement or fear. Rather than attempt to suppress or eliminate the pain, you focus on understanding its characteristics, such as location, intensity, and temperature. This mindful observation helps your brain recognise that these sensations are safe and not threats. As you practice somatic tracking, you may notice changes in the pain, such as a shift in location or intensity. These changes are signs that your brain is adapting to the pain and learning to perceive it as less threatening.

Somatic Tracking: Reframing Pain Perception

Somatic tracking is a mind-body technique that aims to cultivate a non-reactive awareness of physical experiences, allowing us a deeper understanding of how the body communicates discomfort. Somatic tracking helps reframe the perception of pain as a safe sensation rather than a dangerous threat. This shift in mindset can significantly reduce the emotional impact of pain, leading to decreased anxiety and improved coping mechanisms.

Meditation: Calming the Mind and Body

Meditation is not a way of making your mind quiet. It is a way of entering into the quiet that is already there – buried under the 50,000 thoughts the average person thinks every day.
– Deepak Chopra

Meditation, a practice rooted in mindfulness, involves training the mind to focus and redirect attention. It has been shown to effectively reduce stress, anxiety, and pain perception, making it a valuable tool for chronic pain management. During meditation, your focus is on the present moment, observing thoughts and sensations without getting caught up in them. This practice promotes self-awareness and emotional regulation, allowing for a more balanced and resilient response to pain.

I had a vague understanding of meditation but never felt it was for me. How do we know what's for us until we give it a try? Maybe I didn't fully understand what it was all about to look into it deeply. That was until I was in the severest of pain during lockdown, when the world stopped and my mind was racing.

I gave everything a go, and meditation was one of them. When I tried it initially, I was not focusing on what it would do for my mind, and it was very difficult. You don't realise how you have to find a way to get your mind to refocus and reframe the thoughts that keep passing through. My thoughts were scrambling to so many places, and I was struggling to get my mind to focus.

Calming Thoughts

I had never known calming my mind as much as this, and it felt so soothing. I would usually do a range of between five and twenty minutes of guided meditations before I went to sleep to help me get some rest. I would either lie on my bed, which I suppose wasn't always the best way, but I needed to sleep, and it helped. I would also do meditations in a separate room from my bedroom, sitting in a chair or sitting on my bed to feel grounded. The soothing voice and relaxing surroundings were just what I needed. I used to visualise nature and water and feel warm and happy.

I started to feel a deep sense of relaxation and that my nervous system was not in threat mode, like it always was, after my meditation. Yoga Nidra was my starting point before I subscribed to a mobile app called Headspace. I also started to listen to Dr Joe Dispenza, author of *Breaking The Habit of Being Yourself*, who was so on point with his guided meditations. It created space for my thoughts and made me feel emotional. I used to do his morning meditations after doing a little bit of activity. His meditations were twenty minutes in duration. When I was able to, I would sit on the floor cross-legged and focus on the guided meditation, which took my mind to a new dimension.

Being observant of our thoughts and environments is

key to meditation. This enhances our self-awareness when negative, and unhelpful thoughts pop into our minds and can be on a constant loop. I could feel my pain reduce, along with my anxiety, which helped in my recovery. And it created positive emotions.

There is no hard and fast rule as to how long you need to meditate or whether it would work for you. It creates space for your thoughts and allows your nervous system to be calm. I immersed myself in it and felt it was so emotional. I could feel the release of tension from my body. Meditation makes us more self-aware as we focus on the present moment, reducing negative emotions, increasing patience, and reducing stress and anxiety.

Exercise

Body Scan

- Find a quiet space sitting or lying down.
- Close your eyes. Take a few deep breaths through the nose and exhale through the belly.
- Bring your awareness to the top of your head. Notice any sensations or discomfort in your scalp, forehead, or temples. Acknowledge any pain you feel without judgement.

- Move your attention slowly to your face, eyes, cheeks, jaw, and mouth. If there is any pain, inhale a deep breath through the nose and exhale through the belly.
- Move your attention down to your neck and shoulders. If there is any tightness or discomfort, take some deep breaths and just observe it.
- Focus your attention on your chest and upper back. You are not going to try and change the sensations, just observe them, and take some deep breaths.
- Shift your attention to your arms, hands, and fingers. Scan each part while noticing how you are breathing.
- Move your awareness to your lower back, abdomen, and hip. Take a few deep breaths if there is discomfort or painful sensations. Notice what type of sensations they are, and breathe through the discomfort in and out a few times.
- Let your attention travel down to your legs, knees, calves, and feet. Scan each part, and breathe to soothe the sensations you may feel.
- Lastly, take a moment to notice your body as a whole and any changes or pain levels.

You can record the above on your mobile phone, and listen to it if you need it at any time.

Act Now

Massive action, such as moving your body, though it will make you feel uncomfortable will help you through the pain barrier. The brain will be safe in the knowledge that there is no threat and that your body is not damaged. Therefore, activity is a must. It increases blood flow, and the blood circulating the body helps reduce inflammation and stiffness. Blood flow provides essential nutrients and oxygen to the affected areas, and it helps with pain relief and healing. You know how you feel after a workout or walking, where you feel good that you have put your body into action? This is a release of endorphins to get you motivated. How long has it been since you stopped yourself from taking action?

To strengthen my belief in the possibility of recovery, I turned to YouTube videos and discussions featuring Georgie and individuals who had overcome even more challenging diagnoses than mine. Witnessing their complete recoveries and how they were living their best lives inspired me with hope. I immersed myself in recovery stories from the books I read, and even reached out to Steven Ozanich, the author of *The Great Pain Deception*, for support. Steven reassured me that if countless others had successfully recovered, there was no reason why I couldn't do the same. I searched for tangible evidence and personal testimonials, which my unconscious brain needed to accept the possibilities of healing and transformation.

Inaction breeds doubt and fear. Action breeds confidence and courage. If you want to conquer fear, do not sit at home and think about it. Go out and get busy.
– Dale Carnegie

An article by Gretchen Reynolds published in *The New York Times* in November 2021 suggested that tai chi is good for chronic knee pain and to find activities that support you in your healing. Daniel Belavy is a professor of physiotherapy at a university in Germany, and he specialises in how moving influences chronic pain, especially back pain. He stated that "being inactive will tend to reinforce pain sensitivity pathways." It's about finding what works for you, whether that's yoga, swimming, or gardening.[25]

What Steps to Take

What if you're initially fearful as the pain you felt the last time you tried was excruciating and something you want to avoid? What about swimming? This is less impactful on the body and helps strengthen the muscles. Could you aim to do one lap a week? Maybe

25 See: https://www.nytimes.com/2021/11/09/well/move/exercise-chronic-pain.html.

even start by just floating in the water and gradually beginning to move.

Can you walk for two minutes or more? When the pain comes on, are you able to tell your brain you are going to do an extra minute? You can do some calm, deep breathing to self-soothe, which often allows the pain to ease.

Reading Fred Amir's book *Rapid Recovery from Neck and Back Pain* helped me look at what action he undertook and gave me so much hope. Wherever you are on this journey of pain, taking action is a step forward. Take the resistance, fear, and anxiety and place them on the table. Make it a physical thing. Label it. Avoiding the pain will not help in your recovery; it will leave you stuck in the same position as a victim of your thoughts. Emotionally it will leave you feeling low. Leaning into the pain over time will mean that you will start to reap the pleasure. Trust me, it hurts, and it's hard work, but there are so many rewards once you go through it.

"Our deepest fear is not that we are inadequate. Our deepest fear is that we are powerful beyond measure. It is our light, not our darkness, that most frightens us. We ask ourselves, who am I to be brilliant, gorgeous, talented, fabulous? Actually, who are you not to be?"
– Marianne Williamson

Relaxed Mind and Nervous System

As you have started to understand the root cause of your pain, what are you going to do about it? Your mindset and nervous system should be in a place of relaxation if you have understood the previous chapters. And if the pain in your body is now bearable, you can look at taking small steps in whichever way you would like.

You have acknowledged your fears and believe you are not damaged. It is now time to put that work into becoming pain-free on a regular basis. To start, it may be the smallest step. For me, it was walking three minutes non-stop. Remember, it may be painful in the beginning. If you have not exercised some of your muscles for months to years, it could feel uncomfortable. The key thing to remember is to not stop after day one and to keep doing something towards it every day. Keep up the momentum.

I then tried cycling on my exercise bike at home, something that didn't cause as much pain as walking at the time, so I would do it for ten minutes on a high resistance to get the blood pumping. As I gained confidence, I started swimming, which I enjoyed because it felt like I was capable in the swimming pool. When walking up to the swimming pool, I would always be limping. People would ask if I was okay. I feared getting changed as this was difficult for me to do. I encouraged myself to believe that I could do it bit by bit. I started to notice how unfit I was through swimming and how out of breath I would be even in the slow lane. However, even after just a few weeks, I noticed how strengthened my body felt.

Dear overthinker,
You're pacing back and forth, allowing your thoughts to spiral out of control. You're overanalyzing situations, finding more problems instead of solutions. The need for certainty drives you crazy when outcomes aren't in your hands. But in it all, your mind is exhausted and your body is stressed. So, for our own sake, take a step back and breathe deep. Accept what is, release what's not. Have faith that better moments will follow.
– Vex King

Dancing

I love dancing; it's part of my culture. But I was not dancing for joy in my debilitated state. I worried about how I would attend a family party or wedding, and whether I would even be able to sit on a chair, let alone dance. It upset me. I was also relieved to know that it was lockdown, so there were hardly any family gatherings where people could see my sorry state.

I reminisce on the song "Tubthumping" from Chumbawumba with a smile about getting knocked down, but that I will get up again, and I will not be kept down. I used to dance to it at university, and you can tell your brain that too. Give it a go. Search for the song that will energise and motivate you. Do it now!

In Leah Groth's article "9 Reasons Why Dancing Is Good for Your Health (July 2022), she says that dancing boosts self-esteem and psychological wellbeing. However, socially you're also sharing a connection with other people, either in a class, a club, or an environment where you're all enjoying the same thing.[26]

Dance is the hidden language of the soul.
– Martha Graham

We know that dancing engages the whole body, involving cardiovascular health, strength, flexibility, and coordination. I was shocked by how much my mind stretched my body. Dancing releases endorphins, a natural mood-boosting chemical in the brain. It is another way to reduce stress, anxiety, and depression. We feel alive in our minds and bodies. And we feel so good.

---

26 See: https://www.everydayhealth.com/fitness-pictures/health-benefits-of-dance.aspx.

Creating a Habit

To create a habit, starting small to make it easy and achievable is the best way to begin. James Clear makes this a starting point in his book *Atomic Habits*, where momentum is built up daily. Making the habit obvious and visible – for example, having your workout clothes in a place where you can see them as I do – gives visual cues and reminders to make that desired behaviour more tangible. Also, habit stacking, where you pair a new habit with an existing one, increases the likelihood of sticking to your new habit. Lastly, to make it satisfying, reward yourself afterwards.

Writing Down Your Powerful Thoughts

According to Flowly, there are two types of journalling: expressive writing and gratitude journalling.[27]

Victor Frankl, a neurologist and psychiatrist, stated, "Between stimulus and response, there is a space. In that space lies our freedom and our power to choose our response. In our response lies our growth and our happiness."

---

27 See: https://www.flowly.world/post/write-to-heal-science-backed-journalling-to-address-anxiety-chronic-pain-and-more.

Expressive Writing

Expressive writing is all about writing down our thoughts and emotions to move forward. When you are consistent with expressive writing, you create space between yourself and your thoughts in your brain.

Did you know that writing all your thoughts on a piece of paper, a book, a phone, or even doing voice notes is one way of releasing emotional pain? Some of the best songs, poems, and books come from a person's suffering. As the individual goes through the hurt, we feel his or her pain through the lyrics, and the visual creativity in a book or a poem, and it hits our hearts.

The greatest power we have is the
power of our thoughts.
— Marianne Williamson

What is the effect of writing or blurting out all that you are thinking? Well, as you know, you are reading this book to have tools and resources and the hope to heal. You want awareness as to how your emotions cause your chronic pain to continue. According to David Hanscom, we should write down positive and negative emotions, and write for approximately up to twenty minutes per day.

Emotionally regulating yourself, and the nervous

system, is so important. Journalling, let's face it, can sometimes feel like another chore. But you can also see it as your health MOT check, regulating your emotions when you need it, calming your nervous system and your gut, which inevitably will keep your body and mind in check. Nobody sees what you have written, and it can be about anything and anyone, with whatever language you want to use. The best thing is to delete or throw away those written thoughts, and start afresh the next day, so nobody will ever know.

Written expression isn't for everyone. However, if in the past you have had some form of counselling and are still struggling with regulating your emotions, or don't know why you are feeling sad and helpless, and you have not tried writing it down, why not give it a go? Writing in the daytime even just to release a few thoughts may help you notice any repeated thought patterns that keep cropping up and to connect the dots.

The way I hid my trauma behind good behaviour, shyness, and people pleasing as a kid. People thought I was an angel, but really I was terrified to have a personality, so I chose obedience.
– Kaya Nova

When you're in pain, the last thing you want to do is sit there and write and see those dark thoughts that have wreaked havoc in your mind in front of your eyes. It can be scary. If we can allow ourselves to understand that writing thoughts on paper as a physical activity is a huge benefit for our health to heal, mentally and physically, then why not give it a try? If you feel like writing a letter to your future self, this is also a great way to journal. I tried this when my friend asked me to do it a year earlier. She presented the letter on my wedding day a year later, and to say it felt fantastic was an understatement.

You can also write a letter to someone who has impacted your life and for whom you still hold unforgiving thoughts.

Nicole Sachs, a lifelong advocate for chronic pain sufferers through her work as a psychotherapist, speaker, writer, and podcaster, has spoken about how she used to journal. She recommends it for dealing with past stressors such as ACEs, relationships with others that may have significantly affected us, and also present stressors, such as moving house, relationship issues, or health issues. For example, are you good at boundary setting? Are you a people-pleaser, always saying yes to friends or family? But as a result, are you unconsciously angry at people that you know don't serve your best interests or make you feel deflated?

Gratitude Journalling

Gratitude journalling involves just writing three things to be grateful for each day. Gratitude increases our wellbeing, and we start paying more attention to our thoughts as well as understanding more about ourselves. It can help us to release negative emotions of anger or frustration, and to notice something positive in that day.

Robert Emmons, a professor of psychology in California, studied the effects of gratitude on over a thousand people. The participants ranged in age from eight to eighty and were split into two groups. One group was asked to keep a journal in which they were to write five gifts that they were grateful for each day. The other group had to write down five hassles.

Some of the gifts people noted were the generosity of friends and watching a sunset through the clouds. Examples of hassles were things like difficulty in finding a parking space and burning dinner. Emmons found that those who had kept a gratitude journal experienced significant psychological, physical, and social benefits. This included a 25 per cent improvement in overall health and wellbeing in comparison with the group focusing on what had gone wrong each day.[28]

---

28 See: https://www.londonmindful.com/blog/8-wellbeing-benefits-of-practicing-gratitude/.

Gratitude can be seen as a natural antidepressant according to Emily Fletcher, founder of Ziva, a well-known meditation training site. Expressing and receiving gratitude allows the brain to release dopamine and serotonin, which are both important components of our emotions and make us feel good and internally happy. Strengthening the neural pathways every day, like we would a muscle, will help us become more positive.

Exercise

1. Can you start expressive writing about how you feel for a few minutes a day?
2. If question 1 is difficult for you, can you start a gratitude journal and list three things you're grateful for every day, even when you feel like it's been a horrible day?
3. If writing down is something that you do not enjoy, are you able to verbally say what you're grateful for? I do it in the evening whilst I brush my teeth.
4. Could you just ask yourself what three good things happened today? This will help with your emotional regulation.

TAKE BACK CONTROL OF YOUR POWER

Seeing Your Power Grow Again

Neurolinguistic programming (NLP) connects our thoughts, emotions, and physical sensations, harnessing the power of our five senses – vision, hearing, touch, smell, and taste. Through NLP, we can reframe our perceptions of pain. NLP encompasses the conscious and unconscious parts of our brains. The conscious mind is logical and focused on what we already know. The unconscious mind is buried away like little rabbit holes that need a way in.

Anchoring

An important aspect of NLP is the use of anchors, which are specific gestures or touches that bring a desired state of mind, such as confidence or calmness.

Anchors can be enhanced through stacking, using the same or different experiences to cultivate the desired state multiple times. By applying a specific emotional state such as feeling more confident, a neurological association is formed between the stimulus and the emotional state. Anchoring can occur naturally or be intentionally established. They can work as supports in accessing past states and connecting them to the present and future.

I recall attending a training with Toby and Kate McCartney, renowned international trainers, speakers and coaches, in London in 2012. During the training, we focused on the fundamentals of anchoring and how to transform negative states into more positive ones. In my struggle with chronic pain, I began to shape positive thoughts and affirmations in my mind, which I had locked away for a long time. I visualised myself waking up free from pain, vividly imagining my state of being, appearance, and surroundings. I also revisited past positive experiences to bring them into my conscious mind and intensify their realities.

When you visualize, then you materialize. If you've been there in the mind, you'll go there in the body.
– Dr Denis Waitley

Visualisation

Visualisation can effectively reduce stress levels in the body by calming the sympathetic nervous system and decreasing cortisol levels. By engaging all our senses through the imagination, we can achieve a relaxed state of mind, often resulting in a positive experience if we are visual. If you are experiencing chronic pain, you are probably aware of the importance of feeling calm and relaxed. By altering the body's physiology and increasing muscle relaxation, we can start reducing the painful symptoms we continue to experience.

In a 2015 article published in the *Journal of Pain Management Nursing*, studies were shown that demonstrated the significant pain-reducing effects of visualisation on arthritis and other joint issues. This technique is particularly effective for individuals who are visually inclined. If you struggle with picturing images or memories, I suggest exploring alternative senses to recreate a positive state.[29]

Exercise

Visualise the state of how you want to feel, see, or hear. For example, paint a picture in your mind of what being pain-free while walking looks like for you.

Recall images and your senses from your past that you can recreate in your visualisation.

Pay attention to the sounds, the temperature, and any other details you can add.

---

29 See: https://www.arthritis.org/health-wellness/treatment/ complementary-therapies/natural-therapies/guided-imagery-for-arthritis-pain.

FROM PAIN TO POWER

Repeat the process a few times with the same or different positive experiences from the past.

Breathing in Power and Exhaling Pain

Do you know where your diaphragm is located? Well, neither did I until I had to become aware. It lies across the bottom of your lungs and above your stomach. The diaphragm is a muscle. It flattens when you inhale and relaxes when you exhale to allow air in and out of your lungs.

A lot of us use our chests for breathing without any awareness and breathe shallowly and faster than is healthy for us. Diaphragmatic breathing, or calm, slow, deep breathing enhances the intake of oxygen and the release of carbon dioxide. This, in turn, causes the heart rate to slow and can also lower, or stabilise, blood pressure. The vagus nerve, which is the nerve that triggers your body's relaxation response known as the parasympathetic nervous system, is activated while using your diaphragm. It also lowers the stress response system of the sympathetic nervous system.

Breathwork

I first came across how I was breathing when I was being coached by Jeannie Kulwin, who also happened to be a breathwork instructor. She overcame sixteen years of chronic illness from chronic fatigue, back pain, and fibromyalgia. When I asked her what led her to heal, she explained, "Breathwork brought me my power back, and turned off my fear. I was mentally ready given that I had tried lots of things. I heard Steven Ozanich on a podcast, bought his book, and that was such a turning point for me."[30]

Breathwork, breathing slowly, is good for our heart health, nervous system, anxiety, and especially when we are experiencing chronic pain. Conscious breathing exercises can help with relaxation and stress reduction when they involve deep, slow breathing activating the body's relaxation response, and helping to reduce stress and anxiety. We are aware that stress and tension intensify pain, and we need to get some relief where the mind focuses our attention on the present moment, especially on our breathing, and increasing awareness. Did you know that taking slow deep breaths increases oxygen intake and improves our blood circulation? Improved oxygenation can help relax the muscles and alleviate muscle tension.

Box breathing is something I learned while attending peer-to-peer support groups and the online recovery programme SIRPA. This involves taking a slow, deep breath through the nose, counting to four as we inhale, holding our breath for four more seconds,

---

30   Jeannie Kulwin, Stress, Mind-Body and Breathwork Coach.

exhaling through the mouth (or nose) as you again count to four, and then pausing for a count of four before inhaling again. This can be repeated as many times as required.

I used to do this anytime I was feeling severe pain. For example, as I was walking, I started to notice how I was breathing. I would also notice where any anxiety was in my body, which was a tightening of the chest for me. Emotional regulation through box breathing can help calm the nervous system. It can also help manage feelings of anger, frustration, and any other intense emotions.

Exercise

I saw this exercise from the Johns Hopkins All Children's Hospital Department of Anesthesia in America.

- Find a comfortable place and lie on your back.
- Place a hand on your stomach, above your belly button. Place your other hand on your chest.
- Breathe in slowly through your nose and imagine filling a balloon in your stomach. You will feel your hand moving as your stomach gets bigger and pushes out. The

hand on your chest should stay still.

- Breathe out slowly through your mouth and imagine the balloon shrinking as your stomach becomes flat.
- As you breathe in you can imagine a flower, or your favourite food. Breathe in as smoothly and gently as possible.
- As you breathe out, purse your lips and imagine you are gently blowing out birthday candles, or slowly blowing bubbles.[31]

---

31 See: https://www.hopkinsmedicine.org/all-childrens-hospital/ services/anesthesiology/pain-management/complimentary-pain- therapies/diaphragmatic-breathing.

CHAPTER 6

Tracking Progress to Power

Why not track your progress by creating a personalised tracking system for your chronic pain? There's no right or wrong way to do it, so experiment, and find strategies that work for you. Remember, emotion is motion.

The main goal is to consistently motivate yourself towards recovery while acknowledging that chronic pain is *not* a straightforward journey. Pain levels may fluctuate from day to day. They may even increase at times, making you feel like you're back at square one. This is all part of the process. It's important to normalise this experience and not fear it. I have been through it, over it, and under it numerous times. Understand that healing is not linear; it's more like a yo-yo with ups and downs. By regularly tracking and identifying patterns, you'll gain a clearer understanding of your progress.

As part of the SIRPA online recovery programme, I used this statement as my nightly routine. Before going to sleep, write the following statement three times:

> In my dreams, I kindly request the neutralisation of the emotional charge associated with my hip, right leg, and foot. Please utilise my dreams to deactivate the emotions that have been contributing to my symptoms.

A few of my goals for 2021 were to:

1. Stand up straight.
2. Walk without a limp.
3. Pick up my kids without any pain.
4. Run up the stairs.
5. Dance in the morning.

Examples from My Diary – Chronic Pain Tracker

02/07/2020

In my pain journey, I found further hope in the book *The Great Pain Deception* by Steven Ozanich. It sheds light on the connection between his personal life events and the manifestation of pain. After reaching out to Steven and sharing my story, he emphasised the importance of connecting emotions with physical symptoms. He explained that the disc prolapse itself doesn't directly cause pain; rather, healing occurs through deeper self-awareness. This perspective challenged my previous belief that physical actions alone were the key to healing. I knew that while staying active contributes to our overall wellbeing and happiness, true healing requires us to shift our focus from constantly trying to heal and instead embrace the state of *being*. In hindsight, it became clear how this understanding applied to my journey.

15/08/2020

Today was a fabulous day for me as I achieved some significant wins. I triumphantly swam two laps, which was an incredible and invigorating experience. Additionally, I took a thirty-minute walk with my family, making a few stops along the way. Despite these pauses, I recognised that it was perfectly fine and embraced those moments of rest.

Amidst the pain I felt, I made a conscious effort to prioritise self-care by cooking and washing a few dishes. As I encouraged and motivated myself in these everyday activities, I refused to let the pain overpower me. Instead, I aimed to diminish its significance, reducing it to a mere dot. Each day, whether through physical activities or mindful exercises for my mind, I felt the growing strength and resilience within me.

This journey of managing pain has taught me the power of not allowing it to define my every move. It has enabled me to reclaim control over my life and find ways to nurture myself both physically and mentally.

17/08/2020

I walked outside of my house for five minutes, and I spent time playing with the kids, which brought me immense joy. I achieved a significant milestone when I successfully walked to the lamppost at the end of the street and back home without any interruptions. This accomplishment boosted my confidence.

During the day, I stood for six minutes and even managed to cook by intermittently sitting and standing up without letting fear hold me back. I also started brushing my teeth while standing, realising that distracting my mind from the pain momentarily eased it.

As I reflect upon my diary entries, it became evident that simply jotting down my thoughts and feelings after setting a goal each day acted as a powerful motivator, driving me forward, and telling myself, "I've got this."

18/08/2020

After reading Fred Amir's book *Rapid Recovery from Back and Neck Pain,* I found inspiration in journalling my healing journey. I even reached out to Fred Amir for guidance, and coincidentally, he had recently discussed the concept of pain in the brain on a podcast and told me to have a listen.

Today, as I began journalling in the morning, I realised that it seemed to exacerbate my pain, leaving me with mixed emotions. I practised yoga, which was beneficial, but I made the mistake of attempting the McKenzie stretch, previously recommended by my chiropractor, only to aggravate my inner groin area.

During a walk with my daughter, I experienced frustration and halted midway as my thoughts took over. Feeling disheartened, I reached out to yet another chiropractor, which led me to break down and cry. Fear of and obsession with the pain consumed my mind once again. However, I managed to go for a family walk later and achieved ten minutes without stopping. I felt proud of myself, on a high, until a neighbour noticed my struggle as I was misaligned in my walking, and she mentioned a friend who had hip surgery in a helpful manner. The mention of surgery seemed to be a constant buzzword in my ear. I had to continually reinforce in my mind that I would be fine, recognising that stress had amplified my fear, leading to anger and anxiety. I knew that only I could put a stop to my suffering by believing in my healing and visualising it as a reality.

20/09/2020

Throughout the day, I experienced persistent pain in my back and leg, which led me to take two ibuprofen tablets. Surprisingly, just before 6 p.m., I was able to stand for seven minutes! This was a huge improvement. Interestingly, sitting on the soft sofa provided some relief for the pain in my legs. Despite this improvement, I still felt the pain quite intensely, especially since I had woken up in the early hours of the morning in agony.

For most of the day, I spent my time lying on my back and front. I noticed that when I sat up straight, blood flow seemed to increase in my right leg. Walking remained an incredibly difficult task for me. It felt like a constant struggle, and even with ibuprofen tablets, the intensity of the pain persisted.

21/09/2020

Today, I accomplished walking for four minutes not just once but twice. Although there was still pain involved, the experience was still amazing, and I'm determined to continue until it becomes comfortable for me. I managed to stand for six minutes and even washed a few dishes. However, I couldn't find the energy to cook lunch or dinner.

Desperate to find relief from the pain, I tried EFT (emotional freedom technique) based on a friend's recommendation. It involved tapping and proved to be quite helpful in releasing my pent-up emotions. Surprisingly, it also alleviated some of the physical pain as it helped calm my nervous system. Looking back, I can see the extent of my struggles and the lengths I was willing to go through just to find some relief from the pain.

24/09/2020

- Walked comfortably for one minute and rewarded myself with dark chocolate.
- Stood comfortably for one minute and rewarded myself with dark chocolate.
- Managed to stand in the kitchen for one minute and rewarded myself with dark chocolate.

Although standing still posed challenges for me, I pushed through and managed to wash the dishes in one go, even if it took more than five minutes. Sitting on a kitchen chair still sent shockwaves through my nerve endings, but I reminded myself that it was my fear and anxiety that prolonged the pain. I gradually increased the duration of my time sitting on the sofa, training my brain that it was safe and nothing negative would happen.

Reflections

Today marked a significant breakthrough for me. Georgie Oldfield reviewed my MRI results and confirmed that my symptoms were stress-induced, finally putting my mind at ease. It made me question why I needed evidence to validate what she had been telling me all along. Even the McTimoney chiropractor had mentioned that I was experiencing tension in my body. As a sceptic by nature and a lawyer, I relied on evidence to convince my brain that I was safe.

I only needed to lie down once, and I started sitting up on the sofa. It made a remarkable difference as I was able to sit on the kitchen chair. Although I had to move after a while, the fact that I could sit at all made me so happy.

I comfortably walked for one minute followed by an additional two minutes. I also fed my son his lunch and washed up all the dishes while standing. I brushed my teeth while standing comfortably for one minute. Despite the presence of pain, I now acknowledged and accepted that I needed to change my way of thinking.

The coaching session I had today provided much-needed support and attentive listening, which proved to be incredibly important. It helped calm my nervous system and reinforced the belief that I could recover fully by rewiring my neural pathways. I constantly reminded myself there was no threat to my body.

03/10/2020

I included positive affirmations in my recovery such as:

- I can sleep on my side effortlessly for longer than five hours, and I feel amazing.
- I can walk and stand longer each day more effortlessly, and I feel wonderful.
- I can bend over and twist effortlessly, and I feel fantastic.
- I can get out of bed effortlessly like before and stand up, and I feel fantastic.

26/10/2020

I did twenty minutes of yoga for beginners – stretching, ten minutes on the bike, five minutes on level 5 of the bike and then increased to level 6 for a further five minutes. I did two minutes with weights until my baby interrupted me. As I am reflecting on my diary, I started to increase and vary my activities from walking, swimming, yoga, and now weights, which kept increasing my confidence levels.

10/11/2020

I updated my positive affirmations as I was striving for different goals:

- I can walk longer, and I feel really good.
- I can jog a little every day, and I feel amazing.
- I can bend over and twist every day, and I feel fantastic.
- I can be much more active every day, and I feel amazing.

13/11/2020

As I started reflecting on the concept of learning to walk again, it dawned on me how often we simply reach our destinations without pausing to assess how we walked or what our intentions were for the day. We get caught up in the rush, trying to meet deadlines, hurrying the kids along, oblivious to the importance of taking moments to breathe and check in with our bodies. It became clear to me that I had been walking through life without much intention. Despite having ambitious thoughts and desires in my mind, I was still moving at a slow, hunched pace towards my goals. So, what sparked a change for me?

When I faced the reality of mind-body pain, I understood that it was time for a change. I couldn't rely on distractions to escape the pain anymore, I needed to truly listen to my body. Amid heightened anxiety and being in a constant fight-or-flight mode, I began tuning in to the pace of my breath, the beats of my heart, the unsettling churning in my stomach, and the tightening of my chest. It became clear that I was at a crossroads in my life.

Gradually, I started unravelling the complexities of my entire life, going back to my childhood. I opened up my mind and explored my relationship with myself. I delved into the factors that influenced my thoughts and feelings, shedding light on the unconscious patterns that kept me trapped in negativity. As a child suppressing emotions, and coping through being good, I noticed how much unnecessary pressure I placed on myself.

Despite the pain keeping me grounded, I committed myself to walking again. Every day, my goal was to maintain consistency and perseverance, even when I didn't want to make that tiny bit of effort. It took so much out of me. I observed myself with intention, intensity, and purpose, and that's when my mindset began to shift.

I understood that to heal both emotionally and physically, I needed to change. I reminded myself that anyone can embark on this journey of healing with the right amount of willpower. There was nothing to fear.

27/11/2020

- Lifted weights – 5 pounds.
- Biked for twenty minutes.
- Took a short walk.
- Completed reading the book *The Power of Now*, by Eckhart Tolle.

Today marked two months since I started walking continuously without stopping. Ironically, during today's walk, I had to pause due to intense pain. However, unlike before,

I didn't fear it this time. I understood that the pain would subside when the time was right. I learned about the importance of allowing my body to rest and not succumb to the desires of my ego, which constantly pushed me to overdo it. Despite the pain, I'm proud that I still managed to take a walk as well as engage in biking and weightlifting activities.

29/11/2020

- Visualise my movements before doing them pain-free.
- Pay attention to emotions felt in the body.
- Attempt walking.

Today I completed my weightlifting routine and spent twenty minutes on the bike. However, the pain in my left leg intensified, making it difficult for me to walk without limping or having to stop. Deep down, I understand that these sensations are simply my emotions finding a way to express themselves. I also remind myself that this discomfort is temporary and will eventually subside.

01/12/2020

Today, I had an amazing experience by going swimming. It was particularly significant for me as I walked to the leisure centre, which in itself felt like a great accomplishment. Once there, I was able to swim three laps in the slow lane and even changed my clothes standing up, which filled me with a sense of happiness and gratitude to God. I always had faith in God to lift me up in my struggles.

When I went to pick up my daughter from school, I unexpectedly felt a surge of emotions in my stomach, creating a surreal and powerful sensation within me.

31/12/2020

When I initially reached out to Jeannie Kulwin (coach), it was at a time of desperation. I found myself unable to walk or stand for more than two minutes, and a neurosurgeon strongly recommended surgery for my back. Meeting Jeannie on my recovery journey proved to be an incredible blessing. She not only instilled in me the confidence, faith, and trust needed for my healing process but also provided unwavering support every step of the way.

05/01/2021

Despite the heightened pain I was feeling, I walked into the living room for seventeen

FROM PAIN TO POWER

minutes and did a set of weights. Although there was some pain in my left leg, I was determined to overcome it. I even managed to stand and hold my son, and I felt amazing. I even celebrated a small victory by standing up straight in the bathroom.

20/01/2021

Today's workout consisted of various activities. I spent ten minutes on the bike, followed by a session of weights consisting of six sets targeting various muscle groups, including abs. Next, I spent twelve minutes on the treadmill with some holding onto the handlebars for support. Additionally, I dedicated twelve minutes walking in the living room and then stood for ten minutes to work on my balance and stability. I also did a few stretching exercises and took a moment for meditation to focus on mindfulness and relaxation.

05/02/2021

During my workout today, I spent ten minutes on the bike followed by weights. Additionally, I enjoyed two energetic dance sessions to add a bit of fun to my routine. I also went for a twelve-minute jog outside, although I had to make a few stops along the way as the pain became intense. I was left feeling sad and frustrated. I had hoped for more progress, and it was disheartening to face those challenges.

10/02/2021

Today's workout was going to consist of a twenty-minute walk followed by a ten-minute jog. I dedicated fourteen minutes to standing, which is an improvement compared to previous days. I felt so good that the aggravation in my left leg has reduced, which is a significant positive development. Interestingly, I exceeded my initial goal and managed to walk for twenty-two minutes instead of twenty, and I jogged for twelve minutes. This progress has made me feel excited and accomplished. I can even bend down and touch the floor without difficulty. Every day I am getting better, and I am grateful for the progress. Thank you, God!

03/04/2021

I made significant progress on the treadmill today. I accomplished an impressive distance of 4.7 miles, at an incline setting of 11 and maintained it for twenty-five minutes. This achievement left me in awe.

I followed this up by dedicating a further

forty minutes, and noticing that my consistent efforts, no matter how tiny, were improving and strengthening my mind as well as my body. I started to notice that I was writing less in my chronic pain tracker, and I knew there was a reduction in the discomfort I was feeling. It was at this moment that I started forgetting that pain was ever a constant presence in my life. I started to realise that it wasn't merely the physical pain that troubled me but the emotional pain I had been holding onto and never wanted to confront.

This realisation allowed me to understand that true healing goes beyond the physical aspect, and I needed to address the emotional pain that was lingering within.

Evidence to Prove Healing Is Happening

I used to reflect on how far I had come. I wrote down the proof I was progressing as follows:

- I no longer lie on the floor in the middle of the night or day.
- I can comfortably sit on the leather chair in the study room.
- I can stand to brush my teeth.
- I can stand to have a shower.

Exercise

What is your why for recovering from pain? Can you jot down on a piece of paper all the reasons?

What proof do you need to see to keep you on track?

What realistic goals have you set to help your recovery?

Who can support you to keep on track?

The Dopamine Effect

I can give you high blood pressure just on the phone by criticizing you. On the other hand, I can send a tweet to somebody in China and give them a dopamine hit.
– Deepak Chopra

Dopamine allows you to feel pleasure, satisfaction, and motivation, and it acts as a reward system. This is why it felt so good to reward myself after taking baby steps to reach my ultimate goal. Dopamine is a type of monoamine neurotransmitter. It's made in your brain and acts as a chemical messenger, communicating messages between nerve cells in your brain and the rest of your body. For example, if you were to have a pay raise, the brain is activated with a pleasurable reward, which in turn increases dopamine levels.

Lifestyle Changes

Physical self-care is all about engaging in regular movement for your body. A simple walk in nature with natural sunlight for even ten minutes is likely to increase your dopamine levels. Nutrition, such as cutting down on foods and drink containing excessive sugar or salt will also play a part in healing chronic pain. I remember becoming conscious of what I ate or drank when I was in pain, and started having turmeric and sliced lemon in warm water. There is plenty of research online to consider what foods you can become mindful of, especially if you suffer a flare up.

Relaxation is also part of your physical self-care, and I would do this through meditation and getting up early just to have silence in my mind. Physical activity benefits the brain in a positive way and energises us to do more in our day. If you feel achy from a good run or workout and your body feels tense, book a massage to soothe your mind. I tend to go for massages to make me feel good, and it feels so relaxing. It is not to heal the pain.

Dr Anna Lembke, a Stanford psychiatrist and author of *Dopamine Nation*, explores the rise of addiction and its connection to excessive use of technology, social media, and other sources of gratification, and how we can regain control and find a balance for a more fulfilling life.

Sleep

We are already aware of how good sleep is for us. I used to say to others that my body only needed five hours of uninterrupted sleep and I'm okay. When I look back, that wasn't enough for me at all as I was not as focused or sharp as I would have liked to have been, especially at work. I knew my productivity was affected when I had fewer hours of sleep. Now I love getting at least seven hours as I notice how energised my brain and body feel.

Sleep increases dopamine levels, and if we aim to go to bed at a consistent and reasonable time, we will naturally feel well-rested and energised for the day ahead. Dimming the lights before bed and not watching your mobile phone for at least two hours before sleeping will help your mind relax. I do not have caffeine in the late afternoon anymore as I know that can also affect my mind and body, including the body's circadian rhythms. These small changes can improve your cognitive functioning and your mood, helping you to feel refreshed and energised.

Lack of sleep reduces our dopamine levels, and this affects our moods. It may make us feel anxious or stressed. It can also make us feel unmotivated to do anything as we are too tired. Substance abuse, obesity, and stress are all linked to lower levels of dopamine. With low dopamine levels, chronic pain can occur as the combination of lack of sleep, not moving our bodies, and depression is likely to cause pain in the body. Gut issues, weight fluctuations, and fatigue are also effects of having lower dopamine levels.

Powering through the Pain

I began experiencing migraines every third weekend of the month and connected it to long working hours, which left me drained, frustrated, and lacking energy. It started to become evident that my relaxed state triggered these migraines. I wasn't aware I was suffering from chronic stress building within my body. As a lawyer, I was used to a fast-paced legal environment where court deadlines and increased caseloads were familiar. I used to examine medical records, searching for a thorough background to argue that a minor road traffic accident wasn't responsible for the person's psychological or persistent physical pain. What I hadn't realised then was that I was one of those people, with the only distinction being that I never disclosed this to a doctor.

I started neglecting my wellbeing while feeling emotionally responsible for others on my team. I remember returning to work after my father's untimely death, believing that distracting the mind and doing rather than allowing myself to simply be was the practical and beneficial approach.

If you have a high pain threshold, how often have you ignored the sensations that you're feeling and just carried on doing what you were doing?

As the self-proclaimed superwoman I believed myself to be, I found myself suppressing the emotions I was experiencing like I had as a child. I was a first-time mother trying to adapt and balance the demands of a career alongside my role as a parent. No physical injury had occurred for the migraine to have come on, so being mindful of what is happening in the mind and body is important. Our brains will logically conclude lack of sleep and water as the main reasons for the onset of migraines, but there is evidence that shows certain types of people suffer more from such conditions. As we have seen before, this includes our personality traits, such as being people-pleasers and needing to be in control. ACEs can also play a significant role, and in my case led to me being more prone to worry, and feeling anxious.

As a child, starting from the age of five years old, I had the unfortunate firsthand experience of witnessing my mother battling mental illness. This early exposure shaped how I learned to deal with my emotions – by burying them deep down. It became clear to me that I was overly sensitive to others' words, leading to inner turmoil. I couldn't help but obsess over their comments, draining me of energy and focus. I started to understand why I was hypervigilant, and had difficulty regulating my emotions.

It was ironic that the chiropractor I used to visit for my misaligned body used to take two days off a month due to debilitating migraines. If only he was aware. According to an article published by The Association of Migraine Disorders in April 2023, a study involving 1,348 migraine sufferers revealed that 58 per cent of them reported experiencing ACEs. Interestingly, the majority of the sufferers were women, accounting for 88 per cent of the participants. Among the participants, 40 per cent were diagnosed with migraines with auras, and approximately 34 per cent (or about 1 in 3 people) suffered from chronic headaches, experiencing fifteen or more headache days per month.[32]

---

32 See: https://www.migrainedisorders.org/clinical-tips-adverse-childhood-experiences/.

Increasing your self-care and setting better boundaries are examples of altering your personality traits for the better.

Your life does not get better by chance, it gets better by change.
— Jim Rohn

CHAPTER 7

Self-Care to Long-Term Power

Self-care is not selfish. You cannot serve from an empty vessel.
– Eleanor Brown

Self-care is essentially being mindful to look after your mental, physical, and emotional wellbeing. These activities help to increase your self-compassion and self-empowerment. It is not selfish to look after your own needs first, even if you have children. After all, if you are not in the right headspace, it will impact everyone around you.

The five factors of wellbeing are also referred to as the PERMA model, which was developed by the psychologist Martin Seligman, who has written several books on positive psychology.

Positive emotions focus on joy, gratitude, love, and contentment. We need to look for experiences that help bring us happiness and pleasure.

Engagement through our existing hobbies or creating new ones, work, or any other type of activity that creates being in the flow state helps to make time pass quickly and effortlessly.

Relationships through human connection are vital for wellbeing. Maintaining positive relationships with friends, family, colleagues, and community members is likely to bring a sense of belonging, support, and fulfilment.

Meaning is to do with having a purpose in life, aligning those activities with your values, and contributing to the greater good, such as volunteering or through personal hobbies.

Accomplishment is when you have achieved a goal, for example, and celebrate that success, big or small, to help your self-esteem and motivation to keep going.

You, as much as anybody in the entire universe, deserve your love and affection.
– Buddha

Physical Self-Care

Chronic pain can lead to muscle weakness, especially as your flexibility and mobility are significantly reduced. This is usually because we are trying to be careful not to cause more pain to that area, so we avoid mobilising it at all. It is so important to mobilise the areas of pain little by little every day. It might be painful to start with, but after a while, it will become second nature. It will start to reduce your pain sensitivity and likely desensitize the pain signals in the nervous system. Exercise releases endorphins, improving your mood. It can also help to improve your sleep, which allows the body to rest and repair and reduce the pain. You will also feel your power returning as you gain agency over your mind and body. In time, you will be able to independently do the things you want to do once again.

Emotional Self-Care

Emotional care for me is about having time to self-reflect by journalling and meditation, and just having some quiet time by myself. I have to emotionally regulate so that I can understand what is coming up for me if ever I experience discomfort. Deep breathing is another way of calming our nervous system, and I often belly breathe anytime during the day. Life has many curveballs and challenges, and being able to regulate my nervous system and anxiety to help calm it helps me overcome those challenges in a healthier way.

One of my values is connection. I love meeting friends, family, and people in my coaching or networking community to give me a high-energy happy fix. You can find hobbies or activities that bring you joy and make you feel happier. Self-compassion is embedded within my values, and I treat myself more regularly with kindness, which previously was such an alien concept. Our wellbeing is so crucial that we must ensure that we are allowing self-love and attention to be part of this in our daily lives.

Music is the divine way to tell beautiful, poetic things to the heart.
– Pablo Casals

Listening to music may help increase the dopamine levels in our brains, which can uplift people who suffer from depression and attention deficit hyperactivity disorder (ADHD). The mini-ADHD coach highlights that it helps regulate our emotions, especially if the sounds are calming like classical music. I love listening to music, but if the lyrics are heavy, it can transport me back to times that haven't felt so good. I always thought that I couldn't listen to my earphones whilst I was working, yet after the pandemic, as I sat in an office listening to my music, I found I was able to be focused and without any distractions. But music can also distract us. It's just about finding the balance that works for you. Before going to sleep, I now either listen to soothing music or relax with gratitude sayings because I want my mind to be as calm as possible. If I listen to upbeat, loud music, this can make me have trouble sleeping.[33]

Mental Self-Care

Mental self-care is all about engaging in activities that challenge and stimulate the mind. Meditation is one way; this can be guided or silent. Just a few to twenty minutes a day can make a huge difference in observing thoughts and creating space. This helps to reduce anxiety and stress and makes you feel calmer. A 2019 article, "Neurophysical, Cognitive, Behavioural and Neurochemical Effects in Practitioners of Transcendental Meditation", reviewed twenty-one papers confirming that meditation had a positive effect on the brain.[34]

Lifelong learning is another aspect of my self-care. It encompasses activities that increase my intellectual

---

33 See: https://www.theminiadhdcoach.com/living-with-adhd/
adhd-music.
34 See: https://www.scielo.br/j/ramb/a/
BWtGypq4PNSJT4x9kT56zjs/abstract/?lang=en.

curiosity and growth, such as reading books, attending workshops or talks, and listening to my favourite podcasts – such as Dr Rangan Chatterjee – that promote health and wellbeing. Engaging our minds in new and challenging ways strengthens our cognitive functions and memory, and provides a sense of purpose and fulfilment.

Creativity

I love being creative. As a child, I enjoyed doing drawings, painting, and writing poetry as expressions of my feelings. I also enjoyed doing jigsaw puzzles and playing Scrabble. Jigsaw puzzles engage your brain analytically and creatively, exercising the mind and increasing cognition. It's a great way to reduce stress and anxiety. As I grew older, watching my father doing crosswords was something I enjoyed as well. I now enjoy being creative with my children through Play-Doh, LEGO, drawing, and even carving pumpkins.

If the pandemic has taught us anything, it's that we can be creative in many ways, such as cooking, DIY, or anything our minds can focus on. We are aware time away from our screens is healthy, but how many of us actually do it? What about reading a book, if that's your thing, or discovering a new hobby? Or what about going back to a hobby you used to enjoy?

Boundary setting with technology and social media is very important in preventing your mind from becoming overstimulated. Learning to say no when necessary has been difficult but essential for my mental health, especially when it comes to my family. My mental energy is preserved, and I don't feel like burning the candle at both ends.

Social Self-Care

Social self-care is so important for human connection and the positive impact it can have on our overall wellbeing. Having supportive, positive, and fulfilling relationships will foster your happiness. This can include friendships, family, romantic, and professional networks. Spending quality time with one another, doing a shared activity, or just being present with one another is key.

We have spoken about the power of community, and in the struggle with chronic pain, are there positive things you can foster in a peer-to-peer support group or help others who are suffering? Can you join a club or organisation that will bring you together with like-minded people?

Make sure that healthy boundaries are in place and that you are honestly communicating your needs and limits. And honour each other's boundaries. Acts of kindness and compassion through volunteering for a charitable organisation, helping a friend in need, or engaging in random acts of kindness can also improve your wellbeing.

Spiritual Self-Care

Spiritual self-care encompasses the practices that nurture our connections to something greater than ourselves, whether it is a higher power, nature, or a deep sense of purpose within ourselves. It involves engaging in activities that promote inner peace, cultivate meaning, and enhance our overall wellbeing.

Prayer

Prayer has been a cornerstone of spiritual self-care for centuries. It involves expressing gratitude and seeking guidance. Prayer can provide a sense of comfort, faith, hope, and strength during challenging times. For me, prayer has been a daily practice since childhood and has played a significant role in my spiritual growth and resilience. I always have faith that no matter the struggles now, they will ease in time.

Forgiveness: Letting Go of Resentment

Forgiveness is a crucial aspect of spiritual self-care. Holding onto grudges and resentment traps us in a cycle of negativity and hinders our spiritual growth. Learning to forgive ourselves and others, not condoning their actions but releasing them from our emotional grips, can liberate us from the burdens of the past and open us to a more peaceful and fulfilling present. A blog in March 2022, written by Sarah Kircher "self care corner", also mentioned practising forgiveness for self and others as holding grudges uses up our energy and makes us hold on to anger. We want to let go, remain in the present, and look after our wellbeing.[35]

Serving Others: Embracing Compassion

Serving others is a powerful form of spiritual self-care. It allows us to connect with our communities, express compassion, and make positive impacts on the world around us. Whether it's volunteering at a local organisation or charity, helping a neighbour in need, or simply offering a kind word to a stranger, acts of service can bring immense joy and fulfilment to your life as it has for me.

---

35 See: https://sites.psu.edu/shkrcl1/2022/03/15/spiritual-self-care/.

Nurturing Self-Care: Honouring Our Needs

Self-care in general is an essential component of spiritual self-care. When we prioritise our physical, emotional, and mental wellbeing, we create a foundation for spiritual growth. This includes engaging in regular exercise, eating a nutritious diet, getting enough sleep, and practising mindfulness techniques.

Tips for spiritual self-care:

- Spend time in nature.
- Practice mindfulness.
- Do something creative.
- Volunteer your time.
- Take time for yourself.
- Connect with your loved ones.
- Set boundaries.
- Practice self-compassion.
- Seek professional help if needed.

Self care is how you take your power back.
– Lalah Delia

The Power of Values: Guiding Our Choices

Do you know what your values are? A lot of people don't. Not only is the answer to that question personally important, but it is also important professionally. Values act as our internal compasses guiding our decision-making processes. They also help us become aware of what is not working well and what is working well, which aligns with our own beliefs. Just imagine you are working in an organisation, and the values they have in place are not aligned with your own, causing you to feel frustrated and upset and possibly want to quit.

Values help define who we are and what matters most to us, giving our lives a sense of purpose. Our values provide a framework for setting goals, making progress, and ultimately achieving fulfilment. If we consistently demonstrate behaviours that align with our values, we gain trust and respect from others. Values also act as a driving force during challenging times, especially when dealing with difficult decisions, so our motivation and resilience are driven by our values and to make us feel happy.

Safety is not the absence of threat. It is experienced through the presence of connection.
– Gabor Mate

Exercise

List your top twenty values.

See if you can reduce your values to your top five, with your top one being the most important.

Ask yourself whether you are aligned with all your values.

Which values are you in conflict with, and what do you need to do to change that?

The Power Struggle with Your Pain: How to Deal with Setbacks

Anger is remembered pain, fear is anticipated pain, guilt is self directed pain, depression is depletion of energy. Cure – love and joy.
– Deepak Chopra

If you have made progress in your healing journey and feel like you're going back to square one with the onset of pain once again, do not lose hope and sabotage the great work you have done just to come this far. You could have completely recovered only to suffer a setback, weeks, months, or years later. It may be a different area in the body, but the same principles will apply. Don't allow the fear and anxiety to creep back in. Instead, work out what may have caused the setback. Accept the pain, and don't allow your emotions, such as frustration, take over as this will fuel the pain.

Reframe and Reassess

Once you have acknowledged the setback, reframe the situation, and focus on the opportunity for growth once more. Recognise that the pain is temporary. The brain often wants to target areas of our bodies, and you are aware of this now. Allow your mindset to have flexibility. Get that community support to help you through it again, and make sure you keep reaching out. It is more important than ever to see yourself through it. You may think, *Oh, here we go again*, but you will benefit from reminding yourself what you achieved in the past. Use that to continue with it this time. It may feel scary the second time around, but remember that if the pain keeps moving around the body, comes and goes, it's harmless. It may feel scary the second time around, but remember that if the pain is the same as in the past (that was medically checked), or keeps moving around the body, or comes and goes, it's harmless. However, if you are noticing a completely new symptom in your body then you should seek medical evaluation.

Always remember that your present situation is not your final destination. The best is yet to come.
– Zig Ziglar

Keep reminding yourself that any new sensation that is medically checked is safe, and that it's common for the brain to shift pain or illness to another area of your body. Keep telling yourself "This too shall pass". If you have overcome pain in one part of your body, write down, or remember, what steps you took and apply them to any new or reoccurring pain you are feeling. Also, check in with yourself as to what may have caused the setback.

Why Having a Setback May Be Good for Us

Dani Fagan, a fellow hero who had chronic pain, shares ten important lessons on her TMS journey website. She states in one of these lessons that setbacks are normal. Having flare-ups is just an indication that we have an opportunity to keep retraining the brain to get out of chronic pain for good so that it becomes a permanent solution.[36]

Don't get angry at the pain. Remember to treat it like your companion as it's only staying with you for a reason. Label it as something else, or even a favourite

---

36 See: https://mytmsjourney.com/resources/how-to-release-the-fear-of-chronic-pain-and-symptom-relapse/.

cartoon character. It's not trying to hurt you; it wants to protect you and keep you safe. Also, keep asking yourself why the pain has returned. What may have triggered it?

Go back to your why, and give yourself the reasons you can't give up now. What would it mean to you if you became pain-free? What will you do? How will you celebrate?

It's hard, I know, so hard our thoughts are chucking out all kinds of things to stop us in our tracks like we are wading through treacle. We have to block the gremlins. We know they're trying to protect us from the harshness of the world, but we're ready to break through, smash that wall down, brick by brick. If it's uncomfortable or painful now, you don't want to give up. Give yourself a pat on the back, and see how you have progressed. Know that the naysayers will always be there, wanting you to go for another medical consultation, scaring you into believing you are not strong enough to hold on to your power. But you know you are.

Perseverance

You have been ready for a long time; you just needed to be heard and seen with compassion. The inner work can only happen with you. Don't talk about the pain;

think about what you will gain when you become pain-free permanently. Even when one area of your body heals and another pain occurs in your body, you have the tools to overcome it. You know what needs to shift in your mind to do it all over again. The brain is clever and wants to target your body from place to place, and you have all bases covered. You're one step ahead, and you won't fail as you gradually reprogramme your brain with new, healthier neural pathways.

If your progress slows or stops, remind yourself of the tools and strategies you previously used to overcome or relieve the pain. When did you notice pain-free times? Did you track what was happening emotionally at the time? Fear and guilt may have crept back in. They need to be reduced so that you can increase your self-worth and confidence.

When we are looking for hope, for relief from our pain, we go to great lengths to search for many solutions, from listening to others for advice, having ongoing medical appointments, or trying new treatments. However, hopes are often dashed, and it's about powering through the resistance in the brain as much as possible.

Once you reach your destination, you will feel a sense of achievement, empowerment, inspiration, and a transformation so huge. You'll also have an inspiring story to share, and you will want to share it with others. This is no laughing matter. It is a big thing.

Create Your Own Toolbox

Once you have healed, self-care must become non-negotiable. We might not be able to avoid aches and pains in our bodies, but by acknowledging and accepting when they arise through life's curveballs, we can use our understanding and strategies to help ourselves once again.

You will be able to notice through your awareness the sensations, dial it down, and resolve it. This will lead you to becoming aware of what choices you make in life personally and professionally for the betterment of your health and wellbeing.

It doesn't matter if you have had pain for a few months or several years as the same principles apply. There are many recovery stories and much evidence to confirm that you, too, can become pain-free.

Ask yourself what three good things happened in your day. What brings you joy? Start small and re-establish what made you feel happy and gave you that energy to live life.

We know that people-pleasing tendencies are not healthy for us chronic pain sufferers as they blur any boundary setting. Do we even have self-worth if we give in to other people's demands and desires? Do you ever advocate what you want from family and friends? Are you respected enough?

Your value doesn't decrease based on someone's inability to see your worth.
– Unknown

Self-Worth

When we know our self-worth and maintain that we are good enough, that will promote healthy relationships, which remain positive and fulfilling. Boundary setting can happen with ease and no guilt. Those partners and friends treat you with respect and kindness. Self-worth also increases our confidence. I realised that the negative draining energy was not serving me, and I needed to find people who were on an equal footing, without drama, and kind and supportive. It helps to find your "tribe". How does this serve you right now, in the position you are currently in?

According to Sanjana Gupta in "How to Improve Your Self Worth and Why It's Important," self-worth is influenced by a number of factors including childhood experiences, physical appearance, and thoughts and feelings. If we have strong senses of self-worth, we are more likely to persevere through difficulties, believe in ourselves, and foster positive emotions. Asserting who you are is aligned with your values, needs, and aspirations in which your voice is heard. [37]

Dr Romanoff, a clinical psychologist and professor at Yeshiva University, shared some suggestions within this article for things to consider when we think about our self-worth:

Your Self-Care Toolkit:

1. *Recognise your triggers:* Who, or what, is causing your onset of pain or recurrence of flare-ups?

2. *Prioritise your self-care:* Make it non-negotiable and part of your daily routine. It is important to schedule time for yourself and activities that bring you the most joy.

3. *Embrace your self-worth:* Know how much you value yourself, and improve your self-esteem. Set healthy boundaries and assert your needs and wants in relationships.

4. *Celebrate progress:* Acknowledge and celebrate your achievements, no matter how small. Focus on the positive changes you are making in your journey towards your pain.

---

37 https://www.verywellmind.com/what-is-self-worth-6543764

Resources

Quotes:

"Don't get lost in your pain, know that one day your pain will become your cure" (quote taken from a Rumi poem).

"Our deepest fear is not that we are inadequate. Our deepest fear is that we are powerful beyond measure. It is our light, not our darkness, that most frightens us. We ask ourselves, who am I to be brilliant, gorgeous, talented, fabulous? Actually, who are you not to be?" (quote taken from a poem *Our Deepest Fear* by Marianne Williamson).

"Between stimulus and response, there is a space. In that space lies our freedom and our power to choose our response. In our response lies our growth and our happiness." Viktor Frankl – Source unknown

Books

Dr John E Sarno, MD – *Healing Back Pain.*

Georgie Oldfield, MCSP – *Chronic Pain: Your Key to Recovery.*

Howard Schubiner, MD with Michael Betzold – *Unlearn Your Pain.*

Steve Ozanich – *The Great Pain Deception.*

Fred Amir – *Rapid Recovery from Neck and Back Pain.*

David D. Clarke, MD – *They Can't Find Anything Wrong!*

Alan Gordon with Alon Ziv – *The Way Out.*

Bessel Van Der Kolk – *The Body Keeps the Score.*

David Hanscom, MD – *Back in Control.*

Lisa Feldman Barrett – *Seven and a Half Lessons about the Brain.*

Lisa Feldman Barrett – *How Emotions are Made.*

Ruth Kudzi – *How to Feel Better.*

Dr Joe Dispenza – *Breaking the Habit of Being Yourself.*

Gabor Mate – *When The Body Says No.*

Jules Kelly – *The Fading Woman.*

Stephen Covey – *The 7 Habits of Highly Effective People.*

Johann Hari – *Stolen Focus.*

James Clear – *Atomic Habits.*

Nancy Kline – *Time to Think.*

Carol Dweck – *Mindset.*

Eckhart Tolle – *Power of Now.*

Dr Anna Lembke – *Dopamine Nation.*

Mobile Apps

Curable

Calm

Headspace

Podcasts

Dr Rangan Chatterjee – interviewing Howard Schubiner/Dr Joe Dispenza.

YouTube videos and motivational talks; Dan Buglio.

Eddy Lindenstein – eddy@themindandfitnesspodcast. com.

Ruth Kudzi – How to Feel Better. See: https://www. optimuscoachacademy.com/en/how-to-feel-better-podcast/episode-27-chronic-pain-and-coaching-with-narinder-sheena.

Doc Tovah – TMS Roundtable Broadcast on Facebook.

Support

Online Recovery programme – Georgie Oldfield, SIRPA.

Online forums/groups – Living Proof (UK-based), a non for profit organisation founded by Penny George.

Charity groups that help with becoming aware and understanding pain such as Flippin Pain (UK based).

Psychophysiologic Disorders Association (EndChronic Pain.org).

Acknowledgements

I would like to thank my husband Jas for looking after me over lockdown when I was in severe pain while working and looking after our children. He was the bedrock of our family and without his support this book would not have become a reality.

I would like to thank my wider family, and my biggest champion, my mother.

I would like to thank Georgie Oldfield MCSP who had faith and belief in my recovery when I was at my lowest, and providing me with support, evidence, and reassurance that I was going to be okay.

Howard Schubiner for his helpful advice in supporting my recovery and restoring my confidence that I was experiencing a mind-body connection.

Steven Ozanich for providing me with hope for healing.

Jeannie Kulwin you supported my healing and held my hand every step of the way, thank you.

Patricia you really are a great friend and coach and supported me through my pain.

Amanda Chesson for taking your time to review my manuscript, thank you my dear friend.

Ruth Kudzi and Shaa Wasmund without your collaboration, and nurtured support this book would still be a distant memory.

David Clarke for taking the time to support me in endorsing my book and providing helpful recommendations.

About the Author

Narinder Sheena is a passionate advocate for individuals suffering from chronic pain, offering support to those who feel anxious, fearful, and frustrated by their constant discomfort. After transitioning from her previous career as a lawyer, Narinder's life purpose is now focused on helping others achieve a pain-free existence and rediscover the joy in life that they once lost through her coaching and educational talks.

Contact: narinder@mapcoach.co.uk

Printed in Great Britain
by Amazon

44311643R00106